Lecture Notes in Physics

Volume 866

For further volumes:
www.springer.com/series/5304

The Lecture Notes in Physics

The series Lecture Notes in Physics (LNP), founded in 1969, reports new developments in physics research and teaching—quickly and informally, but with a high quality and the explicit aim to summarize and communicate current knowledge in an accessible way. Books published in this series are conceived as bridging material between advanced graduate textbooks and the forefront of research and to serve three purposes:

- to be a compact and modern up-to-date source of reference on a well-defined topic
- to serve as an accessible introduction to the field to postgraduate students and nonspecialist researchers from related areas
- to be a source of advanced teaching material for specialized seminars, courses and schools

Both monographs and multi-author volumes will be considered for publication. Edited volumes should, however, consist of a very limited number of contributions only. Proceedings will not be considered for LNP.

Volumes published in LNP are disseminated both in print and in electronic formats, the electronic archive being available at springerlink.com. The series content is indexed, abstracted and referenced by many abstracting and information services, bibliographic networks, subscription agencies, library networks, and consortia.

Proposals should be sent to a member of the Editorial Board, or directly to the managing editor at Springer:

Christian Caron
Springer Heidelberg
Physics Editorial Department I
Tiergartenstrasse 17
69121 Heidelberg/Germany
christian.caron@springer.com

Dieter Möhl

Stochastic Cooling of Particle Beams

 Springer

Dieter Möhl
CERN
Geneva, Switzerland

ISSN 0075-8450   ISSN 1616-6361 (electronic)
Lecture Notes in Physics
ISBN 978-3-642-34978-2   ISBN 978-3-642-34979-9 (eBook)
DOI 10.1007/978-3-642-34979-9
Springer Heidelberg New York Dordrecht London

Library of Congress Control Number: 2012955941

Printed on acid-free paper

Springer is part of Springer Science+Business Media (www.springer.com)

Preface

The present lecture notes are the transcript from courses which I gave at the CERN Accelerator School. Thanks are due to the different directors of the school for their encouragement to publish these notes and the permission to use material contained in various proceedings. The spirit of these courses and the scope of the transcript are to provide an *introduction* to the principles and the analytical approaches developed to understand stochastic cooling of a particle beam.

The presentation which we have chosen to follow begins with a simplified description which is gradually improved to guide the reader smoothly into the subject. This approach involves a certain amount of repetition which is hoped to be helpful to the newcomer.

Stochastic cooling grew up at CERN. It was fostered in the unique atmosphere of this European laboratory before being taken up by other groups all over the world. I am grateful for the opportunity to report on this enterprise and I hope that the role of those who contributed to this venture is adequately covered in the text and in the references.

Writing up these notes, I could enjoy the hospitality of the CERN accelerator division (PS-Division, subsequently AB- and BE-Department). This also gave me the possibility, to continue the contacts with many outstanding colleagues. I gratefully acknowledge the support.

Geneva   Dieter Möhl

Second Preface—Written by the Author's Colleagues After His Death

The author of this book, Dieter Möhl, was putting the last touches to the manuscript when he died suddenly in May 2012. He was one of the foremost experts on stochastic cooling of particle beams and had assembled material from his contributions to numerous lectures, courses, articles, and conference proceedings into what was to be a unique and authoritative work of reference on this subject

A group of colleagues whom Dieter had asked to comment on his drafts were convinced that the book should not be lost to the accelerator community. The most recent version was complete, requiring only minor textual corrections and adjustments to the table of contents, and the glossary which they had discussed with Dieter just before he died.

With the encouragement of Christian Caron of Springer, we therefore set about the task of preparing this monograph in the series "Lecture Notes in Physics", which fills a gap in the scientific literature. We believe we have preserved the scientific content of the manuscript exactly as Dieter had prepared and we feel honored that we have been able to contribute to Dieter's memory by its publication.

In this we greatly appreciate the help of Dieter's family, in particular of his widow, Dr. Liselotte Möhl, to whom we dedicate our efforts.

CERN

K. Hübner
S. Maury
H. Koziol
L. Thorndahl
E.J.N. Wilson

Contents

Chapter 1
Stochastic Cooling of Particle Beams

1.1 Introduction

Beam cooling aims at reducing the size and the energy spread of a particle beam circulating in a storage ring. This reduction of size should not be accompanied by beam loss; thus the goal is to increase the particle density.

Since the beam size varies with the focusing properties of the storage ring, it is useful to introduce normalised measures of size and density. Such quantities are the (horizontal, vertical and longitudinal) emittances and the phase-space density. For our present purpose they may be regarded as the (squares of the) horizontal and vertical beam diameters, the energy spread, and the density, normalised by the focusing strength and the size of the ring to make them independent of the storage ring properties.

Phase-space density is then a general figure of merit of a particle beam, and cooling improves this figure of merit.

The terms beam temperature and beam cooling have been taken over from the kinetic theory of gases. Imagine a beam of particles going around in a storage ring. Particles will oscillate around the beam centre in much the same way that particles of a hot gas bounce back and forth between the walls of a container. The larger the mean square of the velocity of these oscillations in a beam the larger is the beam size. The mean square velocity spread is used to define the beam temperature in analogy to the temperature of the gas which is determined by the kinetic energy $0.5mv_i^2$ of the molecules.

Why do we want beam cooling? The resultant increase of beam quality is very desirable for at least three reasons:

- *Accumulation of rare particles*: Cooling to make space available so that more beam can be stacked into the same storage ring. The Antiproton Accumulator (AA) at CERN was an example of this (see Fig. 1.1).
- *Improvement of interaction rate and resolution*: Cooling to provide sharply colli-mated and highly mono-energetic beams for precision experiments with colliding beams or beams interacting with fixed targets. The Low Energy Antiproton Ring (LEAR) at CERN is an example of this (see Fig. 1.2).

D. Möhl, *Stochastic Cooling of Particle Beams*, Lecture Notes in Physics 866, DOI 10.1007/978-3-642-34979-9_1, © Springer-Verlag Berlin Heidelberg 2013

Fig. 1.1 The CERN
Antiproton Accumulator
(AA) before 1985. Sketch and
table of performance with
properties of the beam added
every 4.5 s and the stack
accumulated in 24 h (design
values) (By courtesy of
CERN, (©) CERN)

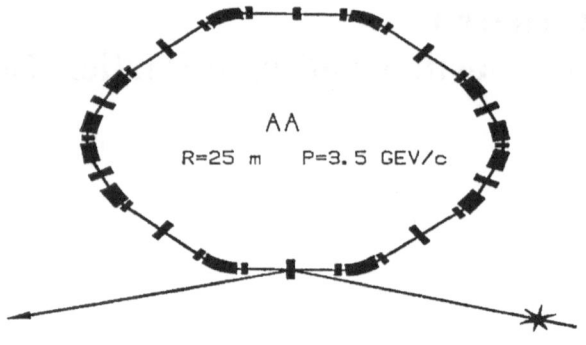

Quantity	Beam in	Stack		Gain
N	10^7	6×10^{11}		6×10^4
E_h	100 π	3.5 π	[mm·mrad]	28
E_v	100 π	2.0 π	[mm·mrad]	50
Δp/p	7.5	2.0	[10^{-3}]	4
$\dfrac{N}{E_h \cdot E_v \cdot \Delta p/p}$	130	4×10^{10}		3×10^8

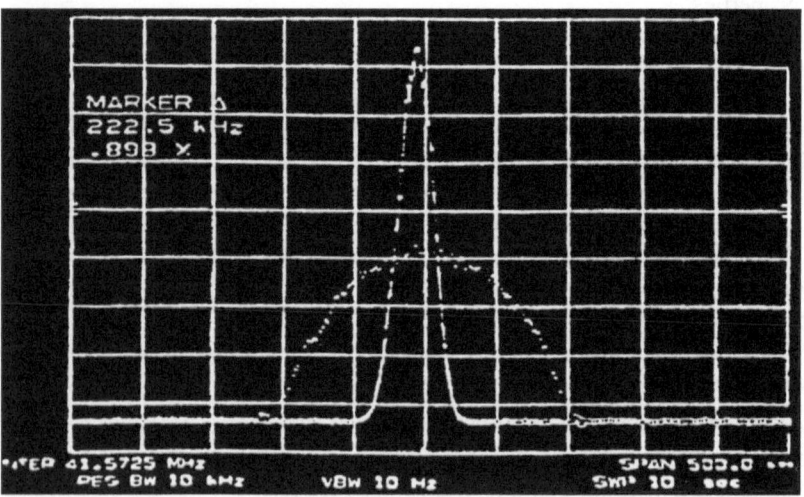

Fig. 1.2 An example (1986) of momentum spread cooling at injection into the Low Energy Antiproton Ring at CERN; the square-root of the density $\sqrt{dN/dp}$ is displayed against frequency; 3×10^9 particles before and after 3 minutes cooling. The momentum spread is reduced by a factor 4 (By courtesy of CERN, (©) CERN)

- *Preservation of beam quality*: Cooling to compensate for various mechanisms leading to growth of beam size and/or loss of stored beam. Again LEAR is an example of this application.

Several cooling techniques have been used in practice or have been discussed [1, 2]: Electron beams have a tendency to cool 'by themselves' owing to the emission of synchrotron radiation as the orbit is curved. The energy radiated decreases very strongly with increasing rest mass of the particles. For (anti-)protons and heavier particles, radiation damping is negligible at energies currently accessible in accelerators. 'Artificial' damping had therefore to be devised, and two such methods have been successfully put to work during the last decade: (i) cooling of heavier particles by the use of an electron beam; and (ii) stochastic cooling by the use of a feedback system.

Chapter 2
Simplified Theory, Time-Domain Picture

2.1 The Basic Set-Up

The arrangement for cooling of the horizontal beam size is sketched in Fig. 2.1. Assume, for the moment, that there is only one particle circulating. Unavoidably, it will have been injected with some small error in position and angle with respect to the ideal orbit (centre of the vacuum chamber). As the focusing system continuously tries to restore the resultant deviation, the particle oscillates around the ideal orbit. Details of these 'betatron oscillations' [3, 4] are given by the focusing structure of the storage ring, namely by the distribution of quadrupoles and gradient magnets (and higher-order 'magnetic lenses') which provide a focusing force proportional to the particle deviation (and to higher-order powers of the deviation).

For the present purpose, we can approximate the betatron oscillation by a purely sinusoidal motion. The cooling system is designed to damp this oscillation. A pickup electrode senses the horizontal position of the particle on each traversal. The error signal—ideally a short pulse with a height proportional to the particle's deviation at the pickup—is amplified in a broad-band amplifier and applied on a kicker which deflects the particle by an angle proportional to its position error.

In the simplest case, the pickup [5] consists of a plate to the left of the beam and a plate to the right of it. If the particle passes to the left, the current induced on the left plate exceeds the current on the right one and vice versa. The difference between the two signals is a measure of the position error. The 'kicker' is, in principle, a similar arrangement of plates on which a transverse electromagnetic field is created which deflects the particle [5].

Since the pickup detects the position and the kicker corrects the angle, their separation is chosen to correspond to a quarter of the betatron oscillation (plus an integer number of half wavelengths if more distance is necessary). A particle passing the pickup at the crest of its oscillation will then cross the kicker with zero position error but with an angle which is proportional to its displacement at the pickup. If the kicker corrects just this angle the particle will from thereon move on the nominal orbit.

D. Möhl, *Stochastic Cooling of Particle Beams*, Lecture Notes in Physics 866,
DOI 10.1007/978-3-642-34979-9_2, © Springer-Verlag Berlin Heidelberg 2013

Fig. 2.1 The principle of (horizontal) stochastic cooling. The pickup measures horizontal deviation and the kicker corrects angular error. They are spaced by a quarter of the betatron wavelength λ_0 (plus multiples of $\lambda_0/2$). A position error at the pickup transforms into an error of angle at the kicker, which is corrected (By courtesy of CERN, (©) CERN)

Fig. 2.2 Representation of the betatron oscillation $x(t)$ of particles; the importance of betatron phase: Particle 1 crosses the pickup with maximum displacement. Its oscillation is (ideally) completely cancelled at the kicker. Particle 2 arrives at an intermediate phase; its oscillation is only partly eliminated. Particle 3 arrives with the most unfavourable phase and is not affected by the system (By courtesy of CERN, (©) CERN)

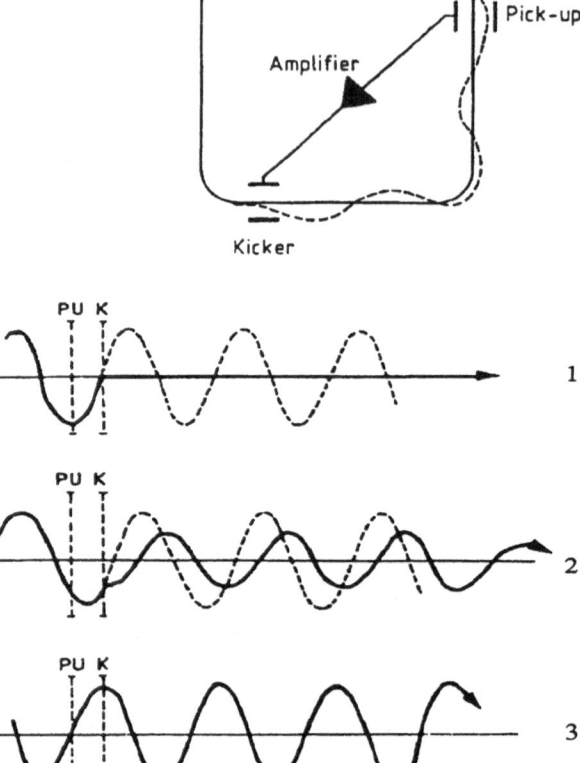

This is the most favourable situation (sketched as case 1 in Figs. 2.2 and 2.3). A particle not crossing the pickup at the crest of its oscillations will receive only a partial correction (cases 2 and 3 in Figs. 2.2 and 2.3). As we shall see later, it will then take several passages to eliminate the oscillation.

Another particularity of stochastic cooling is easily understood from the single particle model (Fig. 2.1): the correction signal has to arrive at the kicker at the same time as the test particle. Since the signal is delayed in the cables and the amplifier, whereas a high-energy particle moves at a speed close to the velocity of light, the cooling path has usually to take a short cut across the ring. Only at low and medium energy ($v/c < 0.5$) is a parallel path feasible.

We have thus familiarised ourselves with two constraints on the distance from the pickup to the kicker. Taken along the beam, this distance is fixed, or rather quantized, owing to the required phase relationship of the betatron oscillation while alternatively, taken along the cooling path, this length is fixed by the required synchronism between particle and signal. A change of the betatron wavelength and/or a change of the energy will therefore require special measures. Incidentally, the first

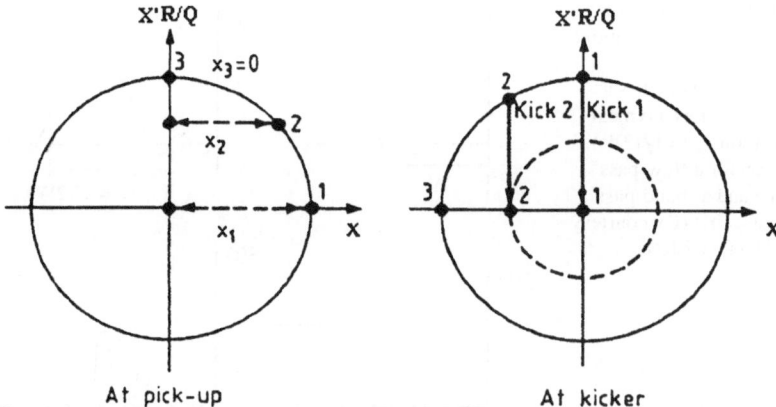

Fig. 2.3 Phase space representation of betatron cooling. The same as for Fig. 2.2, except that a 'polar diagram' is used to represent the betatron motion $x = \tilde{x}\sin\{Q(s/R) + \mu_0\}$, $x' = \tilde{x} \cdot (Q/R) \cdot \cos\{Q(s/R) + \mu_0\}$ (in smooth approximation [5]). The undisturbed motion of a particle is given by a circle with the radius equal to the betatron amplitude $\tilde{x}$. Kicks correspond to a jump of x': the cooling system tries to put particles onto smaller circles. Particles 1, 2 and 3 are sketched with most favourable, intermediate, and least favourable phase, respectively. As the number of oscillations per turn is different from an integer or half-integer, particles come back with different phases on subsequent turns and all particles will be cooled progressively (By courtesy of CERN, (©) CERN)

of these two conditions is due to the oscillatory nature of the betatron motion. For momentum spread cooling in a coasting beam, where the momentum deviation of a particle is constant rather than oscillatory, this constraint does not come into play and a greater freedom in the choice of pickup-to-kicker distance exists.

It is now time to leave the one-particle consideration and turn our attention to a beam of particles which oscillate incoherently i.e. with different amplitudes and with random initial phase. By beam cooling we shall now mean a reduction with time of the amplitude of each individual particle. To understand stochastic cooling, we will next have a closer look at the response of the cooling system. This permits us to discern groups of particles—so called samples—which will receive the same correcting kick during a passage through the system.

2.2 Notion of Beam Samples

To be able to analyse the response of the cooling system, let us start with an excursion into elementary pulse and filtering theory [6, 7]. We will use a bandwidth/pulse-length relation known as the Küpfmüller or Nyquist theorem:[1]

---

[1] The bandwidth/pulse length relation was introduced by Nyquist and independently by Küpfmüller in 1928. This theorem is closely-related to the more general sampling theorem of communication

Fig. 2.4 Illustration of the
Küpfmüller-Nyquist relation:
a signal whose Fourier
decomposition $S(f)$ has a
bandwidth W, has a typical
time duration $T_s = 1/(2W)$.
Illustration for a 'low-pass'
[case (**a**)] and a 'band-pass'
signal [case (**b**)] (By courtesy
of CERN, (©) CERN)

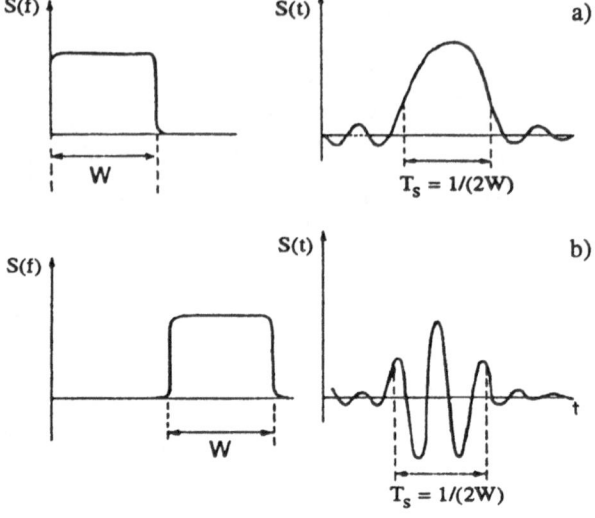

Fig. 2.5 Input and output
signal $S(t)$ of a low-pass
system and 'rectangular'
approximation to the output
pulse $S_f(t)$ (By courtesy of
CERN, (©) CERN)

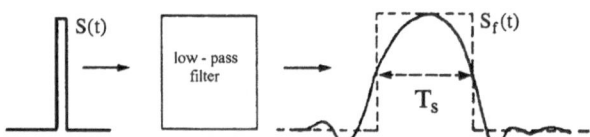

*If a signal has a Fourier decomposition of band-width $\Delta f = W$, then its 'typical'
time duration will be*

$$T_s = 1/(2W)$$

This is illustrated in Fig. 2.4, where we sketch the Fourier spectrum of a pulse
and the resulting time-domain signal. Clearly the two representations are linked by
a Fourier transformation, and this permits us to check the theorem.

For curiosity, note the difference between a pulse with a low-frequency and a
high frequency spectrum (both cases are sketched in Fig. 2.4). In spite of the different shape of the time-domain signal, the 'typical duration' is in both cases $1/(2W)$.
A corollary to the theorem is well known to people who design systems for trans-
mitting short pulses:

*When a short pulse is filtered by a low-pass or band-pass filter of bandwidth W,
the resulting pulse has a 'typical' time width (see Fig. 2.5)*

$$T_s = 1/(2W) \tag{2.1}$$

---

theory which states: If a function $S(t)$ contains no frequencies higher than W cycles per second, it
is completely described by its values $S(mT_s)$ at sampling points spaced by $\Delta t = T_s = 1/2W$ (i.e.
taken at the 'Nyquist rate' $2W$; see, for example, J. Betts [6, 7]).

Table 2.1 An example of samples corresponding to cooling at injection in LEAR [8, 9]	No. of particles in the beam	$N = 10^9$
	Revolution time	$T_{rev} = 0.5\ \mu s$
	Transit time in one pickup	$T_t = 0.1\ ns$
	Cooling system bandwidth	$W = 250\ MHz$
	Sample length	$T_s = 2\ ns$
	No. of samples per turn, n_s	$T_{rev}/T_s = 250$
	No. of particles per sample	$N_s = 4 \times 10^6$

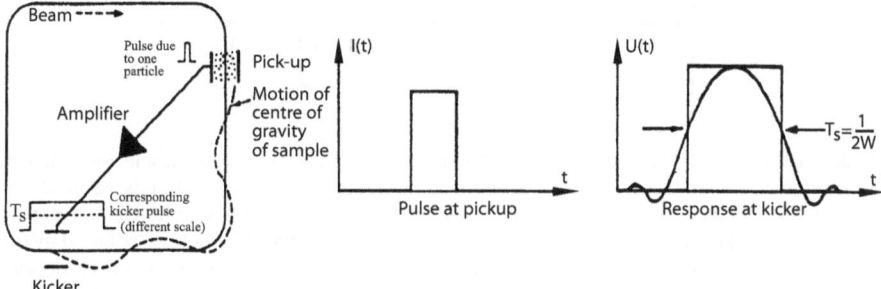

Fig. 2.6 Pickup signal of a particle and corresponding kicker pulse (idealised). The test particle experiences the kicks of all other particles passing within the time distance $-T_s/2 \leq \Delta t \leq T_s/2$ of its arrival at the kicker. These particles are said to belong to the sample of the test particle. Cooling may be discussed in terms of the centre-of-gravity motion of samples (By courtesy of CERN, (©) CERN)

In this form, the theorem is directly applicable to our cooling problem. Passing through the pickup, an off-axis particle induces a short pulse with a length given by the transit time. Owing to the finite bandwidth (W) of the cooling system, the corresponding kicker signal is broadened into a pulse of length T_s. To simplify considerations, we approximate the kicker pulse by a rectangular pulse of total length T_s (Fig. 2.6). A typical set of parameters is given by Table 2.1.

A test particle passing the system at t_0 will then be affected by the kicks due to all particles passing during the time interval $t_0 \pm T_s/2$. These particles are said to belong to the sample of the test particle. In a uniform beam of length T_{rev} (revolution time), there are $n_s = T_{rev}/T_s$ equally spaced samples of length T_s, with

$$N_s = (N/2WT_{rev}) \quad \text{[particles per sample]} \tag{2.2}$$

2.3 Coherent and Incoherent Effects

The model of samples has allowed us to subdivide the beam into a large number of slices which are treated independently of each other by the cooling system. If the bandwidth can be made large enough so that there are no other particles in the

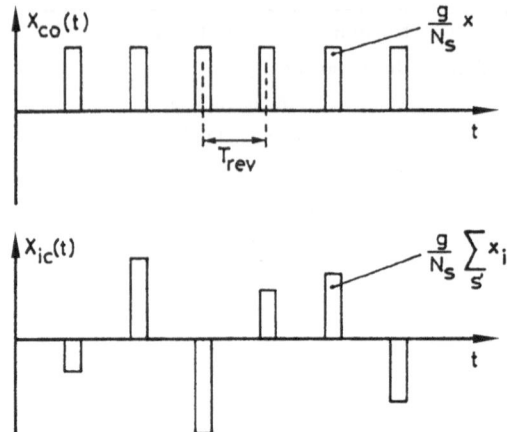

Fig. 2.7 The test particle picture. Signals at the instant of passage of the test particle are sketched. The upper trace gives the coherent correction signal due to the test-particle itself. It is assumed that cooling is slow so that over the turns regarded no reduction of the error is visible. The lower trace sketches the incoherent signal due to the other particles in the sample. The kick experienced by the test particle is the sum of coherent and incoherent effects. If the amplification is not too strong and the sample population is small, the coherent effect which is systematic will predominate over the random heating by the incoherent signals (By courtesy of CERN, (©) CERN)

sample of the test particle, then the single-particle analysis is still valid. However, to account for the reality of some million particles per sample, we have to go a step further and do some simple algebra. This will permit us to discern two slightly different pictures of the cooling process. In the 'test particle picture' we shall view cooling as the competition between: (i) the 'coherent effect' of the test particle upon itself via the cooling loop; and (ii) the 'incoherent effect', i.e. the disturbance to the test particle by the other sample members (see Fig. 2.7). In the 'sampling picture' we shall understand stochastic cooling as a process where samples are taken from the beam at a rate n_s per turn. By measuring and reducing the average sample error, the error of each individual particle will (on the average) slowly decrease.

A few simple equations will illustrate these pictures. Let us denote by x the error of the test particle and assume that the corresponding correction at the kicker is proportional to x, say λx. (Note that the lambda in this context is not that used later for betatron wavelength). With no other particles present, the error would be changed from x to a corrected

$$x_c = x - \lambda x \tag{2.3}$$

i.e. the test particle receives a correcting kick,

$$\Delta x = -\lambda x \tag{2.4}$$

In reality the kicks $-\lambda x_i$ of the other sample members have to be added, and the corrected error after one turn and the corresponding kick are

$$x_c = x - \lambda x - \sum_{s'} \lambda_i x_i$$

$$\Delta x = \underbrace{-\lambda x}_{\text{coherent}} - \underbrace{\sum_{s'} \lambda_i x_i}_{\text{incoherent}} \qquad (2.5)$$

In our rectangular response model, $\lambda_i = \lambda$ is the same for all sample members. Hence, we can also write

$$\Delta x = -\lambda x - \lambda \sum_{s'} x_i \qquad (2.6)$$

Equations (2.5) and (2.6) clearly exhibit the 'coherent' and the 'incoherent' effects mentioned above. The sum labeled s' includes all particles in the sample except the test particle. We may rewrite this sum including the test particle (this sum will be labeled s) and interpret it in terms of the average sample error which is by definition

$$\langle x \rangle_s = \frac{1}{N_s} \sum_s x_i \qquad (2.7)$$

Equations (2.5) and (2.6) then become

$$x_c = x - (\lambda N_s)\langle x_s \rangle \qquad (2.8a)$$

$$\Delta x = -(\lambda N_s)\langle x \rangle_s \equiv -g\langle x \rangle_s \qquad (2.8b)$$

This introduces the second picture. What the cooling system does is to measure the average sample error and to apply a correcting kick, proportional to $(x)_s$ to the test particle. Up to now the sample is defined with respect to a specific test particle; however, to the extent that any beam slice of length T, has the same average error $(x)_s$ our considerations apply to any test particle. This is true on a statistical basis, as will become clear later.

A word about notation: It has become customary to call $g = (\lambda N_s)$, introduced in Eq. (2.8b), 'gain'. However this g is proportional to the amplification (the electronic gain) λ of the system and proportional to N_s. A more precise (but longer) name is 'fraction of observed sample error corrected per turn'.

Now, we can again separate the coherent and incoherent effects and rewrite Eq. (2.6), by using g instead of λ:

$$\Delta x = \underbrace{-\frac{g}{N_s}x}_{\substack{\text{coherent} \\ \text{term} \\ \text{(cooling)}}} \underbrace{- g\langle x \rangle_{s'}}_{\substack{\text{incoherent} \\ \text{term} \\ \text{(heating)}}} \qquad (2.9)$$

Clearly the problem is how to treat the incoherent term. The following approximations will be discussed:

– First approximation: Neglect the incoherent term

- Second approximation: Treat it as a fluctuating random term
- Third approximation: Treat it as a fluctuating random term with some coherence due to imperfect mixing
- Fourth approximation: Include additional coherence due to 'feedback via the beam'

2.4 First Approximation

Neglecting completely the incoherent term in Eq. (2.9) we get a best performance estimate

$$\Delta x = \frac{g}{N_s} x \tag{2.10}$$

We expect an exponential form, $x = x_0 e^{-t/\tau}$ for the error of the test particle which gives the damping rate

$$\frac{1}{\tau} = \frac{1}{x} \frac{dx}{dt} \approx \frac{1}{x} \frac{\Delta x_{per\ turn}}{T_{rev}} \tag{2.11}$$

Substituting into Eq. (2.11) from Eq. (2.10) gives

$$\frac{1}{\tau} = \frac{g}{T_{rev}} N_s \tag{2.12}$$

Interpreting g as the fractional correction, we intuitively accept that it is unhealthy to correct more than the observed sample error, i.e. we assume $g < 1$. Let us put $g = 1$ to make an estimate of the upper limit. Finally it is convenient to express N_s in terms of the total number of particles in the beam and by the system's bandwidth $W : N_s = N T_s / T_{rev} = N/(2W T_{rev})$ [see Eq. (2.2)]. We then obtain, a first useful approximation to the cooling rate:

$$\frac{1}{\tau} = \frac{2W}{N} \tag{2.13}$$

Amazingly enough, this simple relation overestimates the optimum cooling rate by only a factor of 2. However, to gain confidence, we have to justify some of our assumptions, especially the restriction of $g \leq 1$ and the neglect of the incoherent term. In fact, an evaluation of this term will clarify both assumptions and provide guidance on how to include other adverse effects such as amplifier noise.

2.5 Towards a Better Evaluation of the Incoherent Term

To be able to deal with the incoherent term, we make a detour into statistics to recall a few elementary 'sampling relations' [10, 11]. Consider the following problem:

Given a beam of N particles characterised by an average $\langle x \rangle = 0$ and a variance $\langle x^2 \rangle = x_{rms}^2$ of some error quantity x. Suppose we take random samples of N_s particles and do statistics on the sample population—rather than on the whole beam—to determine

(i)   the sample average $\langle x \rangle_s$,
(ii) the sample variance $\langle x^2 \rangle_s$,
(iii) the square of the sample average $[\langle x \rangle_s]^2$, i.e. the square of (i).

What are the most probable values [the expectation values, denoted by $E(\langle x \rangle_s)$, etc.] of these sample characteristics?

For random samples the most probable values are:

(i)   sample average $\rightarrow$ beam average;
(ii) sample variance $\rightarrow$ beam variance;
(iii) square of sample average $\rightarrow$ beam variance/sample population.

Or, in more mathematical language (with $E(x)$ the expectation value of x)

$$E(\langle x \rangle_s) = \langle x \rangle (= 0) \tag{2.14a}$$

$$E(\langle x^2 \rangle_s) = \langle x^2 \rangle = x_{rms}^2 \tag{2.14b}$$

$$E((\langle x_s \rangle)^2) = x_{rms}^2/N_s \tag{2.14c}$$

Results (2.14a) and (2.14b) are in agreement with common sense, which expects that the sample characteristics are true approximations of the corresponding population characteristics. This is the basis for sampling procedures. Equation (2.14c) is more subtle as it specifies the r.m.s. error to be expected when one replaces the population average by the sample average:

$$\langle x \rangle_s = \langle x \rangle \pm x_{rms}/\sqrt{N_s}$$

In words: the larger the beam variance and the smaller the sample size (N_s), the more imprecise is the sampling. In this form, Eqs. (2.14a)–(2.14c) are used in statistics to determine the required sample size for given accuracy and presupposed values for the beam variance x_{rms}^2.

A slightly different interpretation is useful in the present context: suppose we repeat the process of taking beam samples and working our $\langle x \rangle_s$ many times. Although the beam has zero $\langle x \rangle$, the sample average will in general have a finite (positive or negative) $\langle x \rangle_s$. The sequence of sample averages will fluctuate around zero (around $\langle x \rangle$ in general) with a mean-square deviation $\langle x \rangle^2/N_s$. This is the fluctuation (the 'noise') of the sample average due to the finite particle number.

A simple example to illustrate the sampling relations and to familiarise us further with $\langle x^2 \rangle_s$ and $(\langle x \rangle_s)^2$ is given in Table 2.2. It is amusing to note that in this example 'the most probable values' 1/3 and 2/3 respectively [which agree with Eqs. (2.14a)–(2.14c)] never occur for any of the possible samples.

Table 2.2 A simple example of the sampling relations: assume a discrete distribution such that the values $x = -1, 0, 1$ occur with equal probability. Hence the distribution ('beam') average: $\langle x \rangle = 0$, and the distribution ('beam') variance: $\langle x^2 \rangle = 2/3$. Consider samples of size $N_s = 2$. To work out the most probable values of the sample characteristics, write down all possible samples of size $N_s = 2$, determine $\langle x \rangle_s$, $(\langle x \rangle_s)^2$, and $\langle x^2 \rangle_s$ and take the average of these averages to find the expectations. Compare these to the beam averages $\langle x \rangle$ and $\langle x^2 \rangle$

Sequence (sample)		Sample averages		
		$\langle x \rangle_s$	$(\langle x \rangle_s)^2$	$\langle x^2 \rangle_s$
−1	−1	−1	1	1
−1	0	−0.5	0.25	0.5
−1	1	0	0	1
0	−1	−0.5	0.25	0.5
0	0	0	0	0
0	1	0.5	0.25	0.5
1	−1	0	0	1
1	0	0.5	0.25	0.5
1	−1	1	1	1
Expectation (average of column) →		0	1/3	2/3

To conclude our detour, let us mention that the sampling relations (2.14a)–(2.14c) are a consequence of the more general 'central limit theorem' [10, 11] of statistics. For the present purpose we can quote this theorem as follows:

When a large number of random samples of size N_s are taken from a population with statistics $\langle x \rangle = 0$ and $\langle x^2 \rangle = x_{rms}^2$ then the distribution of the sample averages is approximately Gaussian with a mean equal to the population mean and a standard deviation $\sigma = x_{rms}/\sqrt{N_s}$.

2.6 A Second Approximation to the Cooling Rate

Returning to Eq. (2.8a), but re-expressing x_c in full we have,

$$x_c = x - \frac{g}{N_s} \sum_s x_i \tag{2.15}$$

In order to profit from the sampling relations, it is more useful to evaluate the change per turn of $\Delta(x^2) = x_c^2 - x^2$ the change of the squares rather than $\Delta(x) = x_c - x$. We obtain,

$$\Delta(x^2) = -2x \frac{g}{N_s} \sum_s x_i + \left(\frac{g}{N_s} \sum_s x_i \right)^2 \tag{2.16}$$

The second term in Eq. (2.16) gives immediately

$$\left(\frac{g}{N_s}\sum_s x_i\right)^2 = g^2(\langle x\rangle_s)^2 \rightarrow \frac{g^2}{N_s}x_{rms}^2 \tag{2.17}$$

where we have used the sampling relation (2.14c) to express the expected variance of the sample average in terms of the beam variance. To work out the first term we separate the test particle once again from the sum and write

$$x\frac{1}{N_s}\sum_s x_i = \frac{x^2}{N_s} + \frac{x}{N_s}\sum_{s'} x_i \tag{2.18}$$

Next we apply the sampling relation (2.14a) to the remaining sum, i.e. we take

$$E\left(\frac{1}{N_s}\sum_{s'} x_i\right) \approx E\left(\frac{1}{N_s-1}\sum_{s'} x_i\right) = E(\langle x_i\rangle_{s'}) = 0$$

Here we have used $N_s \gg 1$ and assumed that the sample (labelled s') without the test particle is a random sample such that Eq. (2.14a) applies. Thus the expectation value of the first term in Eq. (2.16) is $-2gx^2/N_s$. Clearly this is due to the fact that the x in front 'coheres' with the corresponding term inside the sum. Putting together the terms, the expected change is then

$$\Delta(x^2) \rightarrow -\frac{2g}{N_s}x^2 + \frac{g^2}{N_s}x_{rms}^2 \tag{2.19}$$

Equation (2.19) applies to any test particle. Taking as typical a particle with an error equal to the beam r.m.s. we can write especially:

$$\frac{1}{x_{rms}^2}\Delta(x_{rms}^2) \rightarrow -\frac{1}{N_s}(2g - g^2) \tag{2.20}$$

This gives the cooling rate (per second) for the beam variance:

$$\frac{1}{\tau_{x^2}} = -\frac{1}{T}\frac{\Delta(x_{rms}^2)}{x_{rms}^2} = \frac{1}{N_sT}(2g - g^2) = \frac{2W}{N}(2g - g^2) \tag{2.21}$$

Clearly the term $2g$ presents the coherent effect already identified. The $-g^2$ term represents the incoherent heating by the other particles. The inclusion of this term is the improvement obtained in the statistical evaluation of this section.

It emerges quite naturally from Eq. (2.21) that g should not be too large! In fact, optimum cooling (maximum of $2g - g^2$) is obtained with $g = 1$, and anti-damping occurs if $g > 2$ (see Fig. 2.8). It should be remembered that Eq. (2.21) gives the cooling rate for x^2; the rate $(1/\tau)$ for x is half of this, as can be verified by comparing $x^2 = x_0^2\exp(-t/\tau_{x^2})$ and $x^2 = [x_0\exp(-t/\tau)]^2$.

Fig. 2.8 Cooling or heating
rate when considering the
incoherent term as a random
fluctuation (By courtesy of
CERN, (©) CERN)

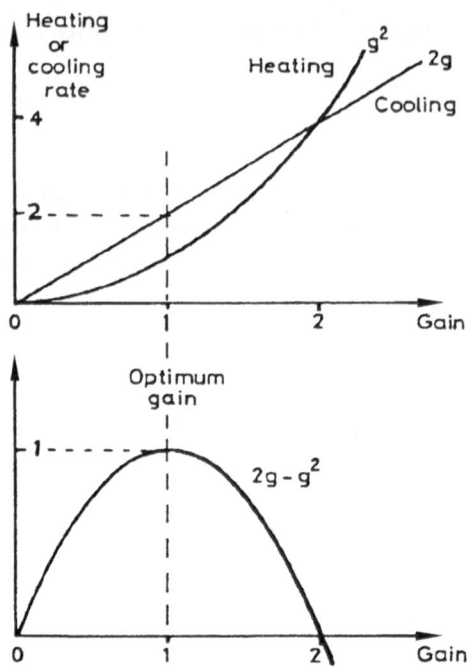

2.7 Alternative Derivation

The way in which we separated the test particle from its sample and regarded the
remainder as a random sample of size $N_s - 1$ may seem unsatisfactory. Let us therefore follow yet another derivation of Eq. (2.21) which is due to Hereward (unpub-
lished notes 1976, see also Ref. [12]). We restart from Eq. (2.16), which we write
as

$$\Delta(x^2) = -2gx\langle x_i\rangle_s + g^2(\langle x_i\rangle_s)^2 \tag{2.22}$$

This is the change for a test particle over one turn. We now take the average of this
over the sample of the test particle (before, we took the average for one particle over
many turns). A slight complication arises from the fact that strictly speaking each
particle defines its own sample, as sketched in Fig. 2.9. We can assume, however,
that the long-term behaviour of any sample (i.e. any beam slice of length T_s) is the
same, so that expectation values are independent of the choice of the sample.

Then the only variable on the r.h.s. of Eq. (2.22) involved in averaging over the
original sample is the x in the first term, and we obtain

$$\langle\Delta(x^2)\rangle_s \rightarrow -2g(\langle x\rangle_s)^2 + g^2(\langle x\rangle_s)^2 \tag{2.23}$$

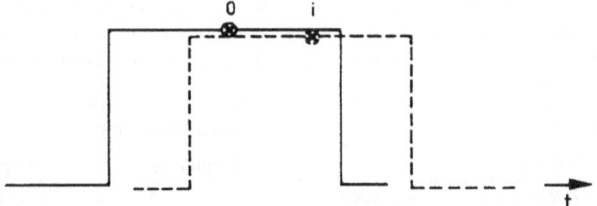

Fig. 2.9 Sample of the original test-particle (0) and of a particle passing earlier (i). Working out the average $\langle x_k \langle x_i \rangle_s \rangle_s$ of $x_k \langle x_i \rangle_s$ each particle has to be associated with its own sample. However, to the extent that all beam samples have the same statistical properties, all long-term averages are the same: $\langle x_k \langle x_i \rangle_s \rangle_s = (\langle x_i \rangle_s)^2 = (\langle x \rangle_s)^2$ (By courtesy of CERN, (©) CERN)

Next we use the sampling relations (2.14b) and (2.14c). We include the fact that the correction (2.23) is applied to all beam samples once per turn. Thus,

$$\langle \Delta(x^2) \rangle_s \to \Delta x_{rms}^2$$

$$(\langle x \rangle_s)^2 \to x_{rms}^2 / N_s$$

and the expected correction of beam variance per turn is

$$\Delta(x_{rms}^2) = -\frac{1}{N_s}(2g - g)x_{rms}^2$$

i.e. exactly as assumed in Eq. (2.20). This leads to the same cooling rate, Eq. (2.21) as obtained by the previous approach.

The derivation sketched in this subsection lends itself to the following formulation of the 'sampling picture'. Take a random beam sample of N_s particles and then correct its average error $\langle x \rangle_s$ by giving a kick $-g \langle x \rangle_s$ to all particles. Owing to the finite particle number of particles in the sample, the beam variance appears as a fluctuation with 'noise' $(\langle x \rangle_s)^2 \to x_{rms}^2 / N_s$ of the centre of gravity. By correcting $\langle x \rangle_s$ to $(1 - g)$ of its value (i.e. to zero for full $g = 1$), one reduces the sample variance (on average) by $(2g - g^2)/N_s$. This should then be repeated N/N_s times per turn and continued for many turns.

Thus, rather than treating single particles, one measures and corrects the centers of gravity of beam samples. It is perhaps not too surprising to note that the total number of measurements, namely the number of turns $n = N_s$ required for reasonable correction multiplied by the number $\ell_s = N/N_s$ of samples per turn, is N, as if we treated the N particles individually.

It is easy to test the sampling prescription for simple distributions; in Table 2.3 we use the previous example (Table 2.2) to verify that the full correction ($g = 1$) reduces the variance by $1/N$, per turn. More generally, the sampling recipe can easily be simulated on a desk computer using a random number generator.

In the next two sections we will use the test particle and the sampling picture alternately to introduce two further ingredients, namely electronic noise of the amplifier and mixing of the samples due to the spread in revolution time.

Table 2.3 'Simulation' of a one-turn correction (with $g = 1$) using the example of Table 2.2. We note down all possible samples of size $N = 2$ and reduce the sample errors to zero by applying the same correction to both sample members. This reduces the beam variance from 2/3 to 1/3, i.e. $\Delta x_{rms}^2 / x_{rms}^2 = 1/N_s = 1/2$

Before correction				After correction			
Sequence		Sample		Sequence		Sample	
		Average $\langle x \rangle_s$	Variance $\langle x^2 \rangle_s$			Average	Variance
-1	-1	-1	1	0	0	0	0
-1	0	-0.5	0.5	-0.5	0.5	0	0.25
-1	1	0	1	-1	1	0	1
0	-1	-0.5	0.5	0.5	0.5	0	0.25
0	0	0	0	0	0	0	0
0	1	0.5	0.5	-0.5	0.5	0	0.25
1	-1	0	1	1	-1	0	1
1	0	0.5	0.5	0.5	-0.5	0	0.25
1	1	1	1	0	0	0	0
Beam variance (average of sample variances)			2/3				1/3

Table 2.4 Signal, noise, and amplification of a cooling system; orders of magnitude for 10^9 particles and 50 s cooling time

Pickup signal	50 nA
Preamplifier noise current	150 nA
Kicker voltage per turn	1 V
Corresponding current (into 200 Ohm)	5 mA
Power amplification	10^{10}

2.8 A Refinement to Include System Noise

A large amplification of the error signals detected by the pickup is necessary to give the required kicks to the beam. Electronic noise in the pickup and preamplifier system then becomes important. In Table 2.4 we anticipate some typical numbers pertaining to transverse cooling of 10^9 antiprotons in LEAR. This example should convince us of the necessity to rewrite the basic equations to include noise. It is convenient [12] to represent noise by an equivalent sample error (denoted by x_n) as referred to at the pickup. We then regard the system sketched in Fig. 2.10 and write

$$x_c = x - g \langle x \rangle_s - g x_n \tag{2.24}$$

Going once again through our basic procedure, taking random noise uncorrelated with the particles we obtain the expected cooling rate

$$\frac{1}{\tau_{x^2}} = \frac{2W}{N} \left[2g - g^2 (1 + U) \right] \tag{2.25}$$

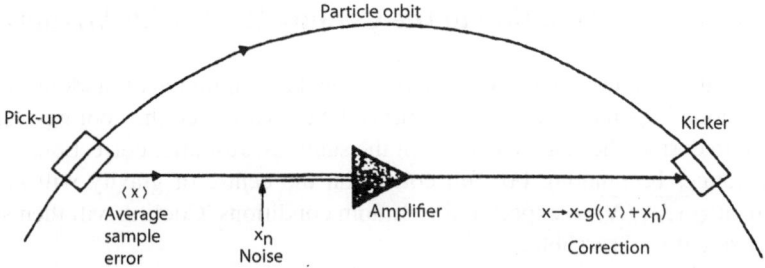

Fig. 2.10 Cooling loop including system noise. The noise is represented as an equivalent sample error $x_n(t)$ as observed at the pickup (By courtesy of CERN, (©) CERN)

Here $U = E(x_n^2)/E(\langle x \rangle_s^2)$ is the ratio of the expected noise to the expected signal power, called the 'noise-to-signal power ratio' or 'noise-to-signal ratio' for brevity. We should note here that for ions $U = U_1/Z^2$ where Z is the charge number of the particle (the number of electrons stripped off) and U_1 the noise-to-signal ratio calculated for singly charged particles. This introduces the noise into our picture: it increases the incoherent term by $(1 + U)$. System noise and the disturbance caused by the other particles enter in much the same way; the latter is therefore also called particle noise.

Several things can be observed from Eq. (2.25): Cooling remains possible despite very poor signal-to-noise ratios ($1/U \ll 1$). All we have to do is to choose g small enough ($g < g_0 = 1/(1 + U) \approx 1/U$), which unavoidably means slow cooling ($\tau_{x^2} > NU/(2W)$). In other words, we have to be patient and give the system a chance to distil a signal out of the noise. In the 1978 Initial Cooling Experiment (ICE) [13] with 200 circulating antiprotons the system worked with signal-to-noise ratios as low as 10^{-6}.

Secondly, U has a tendency to increase as cooling proceeds: the electronic noise tends to remain the same, whereas the signal decreases as the beam shrinks. This is the case for transverse cooling unless the pickup plates are mechanically moved to stay close to the beam edge as was done in the Antiproton Collector [14, 15] at CERN.

With changing U, cooling is no longer exponential. Equation (2.25) gives the instantaneous rate, and cooling stops completely when U has increased such that $1 + U = 2/g$. In this situation, equilibrium is reached between heating by noise and the damping effect of the system. To avoid this 'saturation' it is sometimes advantageous to decrease g during cooling in order to work always close to the optimum gain [maximum of Eq. (2.25)] $g = g_0 = 1/(1 + U)$. In all cases it is important to obtain a good signal-to-noise ratio. Frequently, this means having a large number of pickups as close as possible to the beam, as well as high quality, low-noise preamplifiers often working at cryogenic temperatures.

2.9 Third Approximation to the Cooling Rate (with Mixing)

So far, all our considerations have been based on the assumption of random samples. This is a good hypothesis for an undisturbed beam. However, the cooling system is designed to correct the statistical error of the samples. Just after correction, samples will no longer be random. For full correction the centre of gravity will be zero, rather than $\sqrt{x_{rms}^2/N_s}$ as expected for random conditions. Cooling will then stop as no error signal is observable.

Fortunately, owing to momentum spread, particles in a storage ring go round at slightly different speeds, and the faster ones continuously overtake the slower ones. Because of this mixing, the sample population changes and the sample error reappears, until ideally all particles have zero error. The dispersion of revolution time with momentum is governed by

$$\frac{\Delta T_{rev}}{T_{rev}} = \eta \frac{\Delta p}{p} \qquad (2.26)$$

where the off-momentum function (also called "slip factor") [3, 4]: $\eta = \gamma_t^{-2} - \gamma^{-2}$ is given by the distance of the working energy (γ) from transition energy (γ_t).

If mixing is fast so that complete re-randomization has occurred on the way from the kicker to the pickup then the assumption of random samples made in the previous sections is valid. If however, mixing is incomplete, cooling is slower. In fact, if it takes M turns for a particle of typical momentum error to move by one sample length with respect to the nominal particle ($\Delta p/p = 0$), then intuitively one expects an M times slower cooling rate.

A slightly different way of looking at imperfect re-randomization suggests itself in the frame of the test particle picture: bad mixing means that a particle stays too long—namely M turns rather than 1 turn—together with the same noisy neighbours. This increases the incoherent heating by the other particles by a factor M. We thus generalise the basic Eq. (2.25) (a more rigorous derivation will be given later in the time-domain analysis)

$$\frac{1}{\tau_{x^2}} = \frac{2W}{N}\left[2g - g^2(M + U)\right] \qquad (2.27)$$

and call $M \geq 1$ the mixing factor. It is defined as the number of turns for a particle with one standard deviation in momentum to migrate by one sample length T_s. Equation (2.27) has the optimum

$$\tau_{x^2} = (N/2W)(M + U) \quad \text{for } g = g_0 = 1/(M + U) \qquad (2.28)$$

This underlines the importance of having good mixing ($M \to 1$) on the way from correction to the next observation. However what about mixing between observation and correction? Surely if the sample as observed is very different from the sample as corrected, then adverse effects can happen. Let us again resort to the test particle description and try to imagine how the coherent and the incoherent effects change.

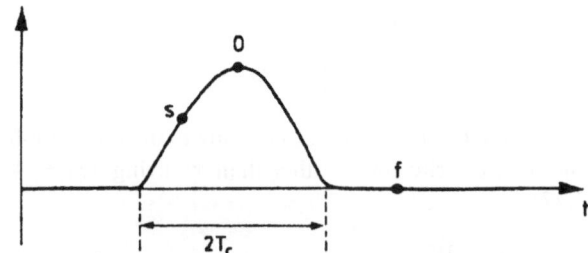

Fig. 2.11 Synchronism between particles and their correcting pulse on their way from pickup to kicker. The response of the cooling system to a particle (the 'coherent effect') is approximated by a 'parabola' $s(t) = 1 - (\Delta t/T_c)^2$ of width $\pm T_c$ instead of the 'rectangle' used in Figs. 2.5 and 2.6. A nominal particle (0) arrives at the kicker simultaneously with the correction kick. The particle f is much too fast and advances its correction pulse. The particle s is slightly too slow. Thus, the three particles receive full correction, no correction, or partial correction, respectively (By courtesy of CERN, (©) CERN)

As to the latter, we expect that it is to first order not affected. We can just assume that the perturbing kicks are due to a new sample which has the same statistical properties as the original beam 'slice'. The coherent effect will, however, change because the system will be adjusted in such a way, that the correction pulse will be synchronous with the nominal particle ($\delta p/p = 0$). Particles that are too slow or too fast on the way from pickup to kicker will therefore slip with respect to their self-induced correction (Fig. 2.11). In fact, in the rectangular response model used above (Figs. 2.5 and 2.6), the coherent effect will be completely zero if the particle slips by more than half the sample length.

At this stage, it is more realistic to use a parabolic response model of the form $1 - (\Delta t/T_c)^2$. Here Δt is the time-of-flight error of the particle between pickup and kicker; T_c, the useful width of the correction pulse, is about equal to the sample length T_s for a low-pass system. But T_c is shorter than T_s for a high-frequency band-pass system, with a response as sketched in Fig. 2.4(b). This is because for the coherent effect, only the range from the maximum to the first zero crossing right and left in Fig. 2.4 is useful. Introducing the typical error $\Delta t_{PK} = t_{PK}\eta_{PK}\Delta p/p$ and calling $\Delta t_{PK}/T_c = 1/\tilde{M}$ we can modify the coherent term $g \to g(1 - \tilde{M}^{-2})$ to account for unwanted mixing between observation and correction. Here η_{PK} is the 'local off-momentum function (local phase slip factor) pickup to kicker'.

For transverse cooling, where momentum error and position error are uncorrelated, one can simply insert as Δt_{PK} the r.m.s. error, to obtain the beam cooling time. For the longitudinal case the r.m.s. cooling time involves averaging over $(\delta p/p)^4$ which depends on the momentum distribution. Then $\tilde{M}$ becomes more involved but the simple approximation $g \to g(1 - \tilde{M}^{-2})$ will be retained here just as an orientation.

In a regular lattice the flight time from pickup to kicker is a fixed fraction of the time from kicker to pickup, and the two mixing factors M and $\tilde{M}$ are proportional to each other, $M = \alpha_M \tilde{M}$, with $\alpha_M \approx \alpha_T$ given by the ratio of the corresponding flight times.

The modification of g in the coherent term

$$g \rightarrow g\left[1 - \tilde{M}^{-2}\right] \tag{2.29}$$

leads to a final form of the basic equation. Returning to the r.m.s. cooling rate which, following convention, we write for x rather than x^2 using $1/\tau = \frac{1}{2}/\tau_{x^2}$, we have instead of Eq. (2.27)

$$\frac{1}{\tau} = \frac{W}{N}\left[2g\left(1 - \tilde{M}^{-2}\right) - g^2(M + U)\right] \tag{2.30}$$

Equation (2.30) is the main result of our simple analysis. It exhibits some of the fundamental limitations of stochastic cooling.

By a clever choice of the bending and focusing properties of the storage ring it is possible, in principle, to make $\eta_{PK} = 0 \rightarrow \Delta t_{PK} = 0$, and Δt_{KP} large, to approach the desired situation of $\tilde{M}^{-2} = 0$ and $M = 1$ [16]. But this complicates the storage ring lattice. The compromise adopted in existing rings is to sacrifice some of the desired re-randomisation in order to avoid too much unwanted mixing.

We note that $1/\tau$, Eq. (2.30) has a maximum

$$\frac{1}{\tau_0} = \frac{W}{N}\left[\frac{(1 - \tilde{M}^{-2})^2}{M + U}\right], \tag{2.31}$$

characterised by

$$g = g_0 = \frac{1 - \tilde{M}^{-2}}{M + U} \tag{2.32}$$

Then, in the best of all cases ($M = 1$, $U = 0$, $\tilde{M}^{-2} = 0$) Eq. (2.31) yields $\tau = N/W$. As an example, to fix ideas, we take $W = 1$ GHz and obtain $\tau = 1$ s at 10^9 or $\tau \approx 1$ day at 4×10^{13} particles.

To include mixing, we assume that the time-of-flight dispersion between pickup and kicker and between kicker and pickup are such that the unwanted mixing is one half of the wanted mixing, i.e. we put (as an example) $\tilde{M} = 2M$. We further assume that the sensitivity and the number of pickup units is such that $U = 1$. Then [from Eq. (2.31)] the fastest cooling, obtained with $M \approx 1.4$, is $\tau = 3.15N/W$. As a further example we take again the mixing conditions $\tilde{M} = 2M$ but assume large noise $U = 10$. Then with $M \approx 2.3$ we obtain the best $\tau = 13N/W$. We retain that over a wide range of parameters τ is proportional N/W. From Fig. 2.12 we conclude that 'classical' cooling systems follow a 'working line' where the proportionality constant is about 10. The 'leveling off' of the τ vs. N curves in Fig. 2.12 for small N is due to the fact that the noise U is proportional to $1/N$.

As to the bandwidth: 250 to 500 MHz was standard in the 'first generation cooling experiments'; 4 GHz came later and bands up to 8 GHz were tried (without full success) in the CERN-ACOL and the Fermilab antiproton sources. Frequencies of 10 GHz have been contemplated in the literature [16]. More recent proposals of 'optical stochastic cooling' [17, 18] discuss bands of 10^{13}–10^{14} Hz but then

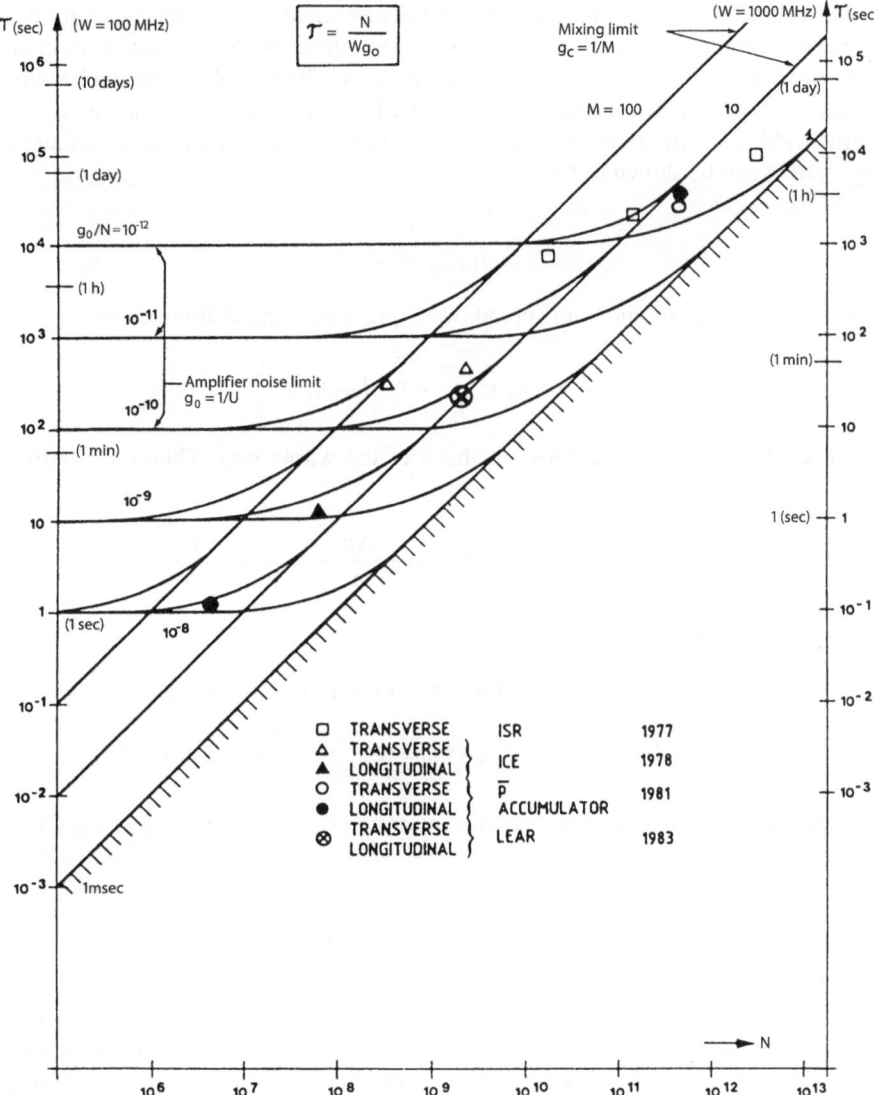

Fig. 2.12 Cooling time versus intensity. The *vertical scale* is normalised for 100 MHz (*left*) and 1000 MHz (*right*) bandwidth. The *inclined lines* represent the mixing limit, the *hatched* one gives the lowest possible $\tau \approx N/W$. For low intensity the cooling time levels off because of noise (noise limit). The points represent initial cooling in various machines. These points roughly follow a line with $\tau \approx 0.10N/W$. During cooling, noise and/or mixing tend to become more important and the cooling time longer (By courtesy of CERN, (©) CERN)

the problem of mixing between observation and correction requires a time spread and stability $\Delta T \ll 10^{-13}$–10^{-14} s. Even with special lattice insertions this is a formidable task.

To conclude this subsection we remark the unwanted mixing imposes a limit on the bandwidth or, more precisely, on the upper frequency of the cooling band. In fact if the time of flight error δt_{PK} of a particle is bigger than T_c, then cooling becomes ineffective for this particle. For a band-pass with flat response from $f_{\min}$ to $f_{\max}$ the useful width $\pm T_c$ [first zero crossing of $S(t)$ right and left from the maximum in Fig. 2.4(b)] can be shown to be

$$T_c \approx \frac{1}{2(f_{\min} + f_{\max})}$$

To relate this to the momentum spread, we express the time of flight error:

$$\delta t_{PK} = t_{Pk}\eta_{PK}\frac{\Delta p}{p} = \alpha_T T_{rev}\eta_{PK}\frac{\Delta p}{p}.$$

For a regular lattice η_{PK} is close to the η of the whole ring. Then the condition $\delta t_{PK} < T_c$ yields

$$\frac{1}{2(f_{\min} + f_{\max})} > \alpha_T T_{rev}\eta_{PK}\frac{\Delta p}{p} \approx \alpha_T T_{rev}\eta\frac{\Delta p}{p}$$

Taking as an example:

$$\alpha_T = 0.5 \quad \text{(flight time: pickup to kicker/circumference)},$$

$$f_{rev} = 1.5\,\text{MHz}, \qquad \eta_{PK} \approx \eta = 0.1, \qquad \frac{\Delta p}{p} = 2 \times 10^{-2}$$

(corresponding in round numbers to the Antiproton Accumulator Ring AA at CERN in 1984) we obtain

$$f_{\min} + f_{\max} \approx f_{\max} < 0.75\,\text{GHz}$$

This is close to the band limit used for the AA precooling system (0.5 GHz); although the limit is not strictly applicable to this system because the different cooling systems cover only parts of the aperture and do not 'see' the full $\Delta p/p$. In any case a small value of $\alpha_T T\eta_{PK}$ (short distance pickup to kicker, small ring, operation close to transition energy and/or special optics) is essential to be able to work with large bandwidth. Later we will also see, that for momentum cooling by the filter method even more restrictive mixing limits pertain.

2.10 Practical Details

So far we have, in a general way, discussed a system for correcting 'some error x'. In practice cooling is used to reduce the horizontal and/or vertical betatron oscillation and the momentum spread of the beam. Table 2.5 gives a summary of the corresponding hardware.

Table 2.5 Stochastic cooling systems in use or proposed

Type	Pickup	Corrector	Optimum pickup to kicker distance
Betatron cooling	Difference pickup	Transverse kicker	$(2k+1)\lambda_\beta/4$
Momentum cooling, Palmer-Hereward method	Difference pickup	Longitudinal kicker	$(2k+1)\lambda_\beta/2$ (for Hereward cooling)
Momentum cooling, filter method	Longitudinal (sum) pickup + comb-filter	Longitudinal kicker	–
Momentum cooling, transit time method	Longitudinal pickup + differentiator or two longitudinal pickups	Longitudinal kicker	–

The simple time-domain approach can be directly applied to momentum cooling by the Palmer method. This will be done in the next subsection. A discussion of betatron oscillation cooling will follow in Sect. 2.12. Simultaneous momentum and betatron cooling by Hereward's method will be treated in Sect. 2.13. Momentum cooling by the filter and transit time method is more naturally discussed in frequency-domain, Chap. 4. A simple time-domain description is sketched in Sect. 2.14.

2.11 Momentum Cooling (Palmer Method)

A horizontal position pickup is used to detect the horizontal orbit displacement of a test particle $x_p = D_{PU}(\delta p/p)$ concurrent with its momentum error; D_{PU} is the value of the "orbit dispersion function" at the pickup as determined by the focusing properties of the storage ring. In addition to the momentum dependent displacement there are further contributions to the position error, especially the betatron oscillation (x_β) of the particles. We shall neglect this contribution here, assuming that the pickup is placed in a region of large dispersion so that x_p dominates over x_β. We are then in a situation where momentum cooling as envisaged by R. Palmer (private communication to L. Thorndahl and H.G. Hereward in 1975) is possible. At the RF gap the particle receives a longitudinal 'kick' hence a change of momentum proportional to the detected error.

The basic one-passage equation for $x = D(\delta p/p)$ of a test-particle (including noise) is

$$x_c = x - g\big(D\langle\delta p/p\rangle_s + x_n\big) \tag{2.33}$$

This is completely equivalent to Eq. (2.15), thus leading to the cooling rate (2.30) for momentum deviation.

Above we tacitly implied that at the kicker the dispersion function D_K as well as its derivative, D'_K are zero. Otherwise the momentum correction leads to an excitation of betatron oscillations. The reason is that the momentum kick introduces an abrupt change of the equilibrium orbit and the particle starts to oscillate around this new displaced orbit. This is an example of 'mutual heating', a problem common to most cooling systems.

The more general case where both x_p and x_β are present at the pickup and where D is non-zero at the kicker was analysed by Hereward [12]. He showed the mutual heating and at the same time the possibility of using the Palmer system for simultaneous longitudinal and transverse cooling by a suitable choice of the pickup to kicker distance [12, 19]. We will sketch his derivation (Sect. 2.13) after having dealt with betatron oscillation cooling.

2.12 Betatron Oscillation Cooling

The system to be discussed (sketched in Fig. 2.1) is the one considered by van der Meer in the first paper on stochastic cooling [20]. The single passage correction of the position (x) of a test particle at the kicker can be described by:

$$x_c = x$$
$$\beta_k x_c' = \beta_k x' - g \langle x_i \rangle_{s,PU} \tag{2.34}$$

i.e. its position remains unchanged but its orbit angle is corrected in proportion to the sample error at the pickup. Here, the beta function (β_k) at the kicker has been introduced so that g is dimensionless as before, $\langle x_i \rangle_{s,PU}$ is the position error at the pickup of the sample belonging to the test particle. We will denote $\langle x_i \rangle_{s,PU} \equiv x_{s,PU}$ in the following

To ease the discussion one can use the smooth sinusoidal approximation of the betatron oscillation, but it is simple enough to retain the exact equations which include the variation of the lattice functions around the ring. A way to proceed is to calculate the long-term change of the "invariant" ε, which represents the square of the betatron amplitude normalised by the beta function, or equivalently: the phase-space area (also called "Courant and Snyder invariant" or "single particle emittance" [3, 4]) enclosed by the motion of the particle. In fact the betatron motion of any particle can be written [3, 4] as

$$x = (\varepsilon\beta)^{1/2} \cos\psi$$
$$x' = -(\varepsilon/\beta)^{1/2}(\alpha\cos\psi + \sin\psi) \tag{2.35}$$

Here $\psi = \int ds/\beta$ is phase of the betatron oscillation, $\beta = \beta_x$ and $\alpha = -\beta'/2$ are the familiar Twiss parameters [3, 4]. We omit the subscript to the beta function here to the extent that no confusion with the relativistic factor is possible. The invariant is written as

$$\varepsilon = \beta^{-1}\left[x^2 + (x'\beta + \alpha x)^2\right] \tag{2.36}$$

From Eqs. (2.34) and (2.36) one has the corrected value after one passage

$$\varepsilon_c = \beta_K^{-1}\left[x^2 + \left(\beta_K x' - g x_{s,PU} + \alpha_K x\right)^2\right] \tag{2.37}$$

Here the Twiss functions α_K and β_K are taken at the kicker. The pickup is at a phase advance $-\Delta\psi$ upstream of the kicker. Let the test particle pass the kicker with phase ψ. Then its position at the pickup was

$$x_{PU} = (\varepsilon\beta_{PU})^{1/2}\cos(\psi - \Delta\psi) \tag{2.38}$$

Working out the change $\Delta\varepsilon = \varepsilon_c - \varepsilon$ from Eqs. (2.35) and (2.37) we have a linear and a quadratic term in $g x_{s,PU}$. As to the first we retain, once again, only the contribution of the test particle and find for the long-term expectation value:

$$-2g x_{s,PU}\left(\beta_K x' + \alpha_K x\right) \rightarrow -2g\frac{x_{PU}}{N_s}\left(\beta_K x' + \alpha_K x\right) \rightarrow -g\frac{\varepsilon}{N_s}\beta_K^{1/2}\beta_{PU}^{1/2}\sin(\Delta\psi) \tag{2.39}$$

Here Eqs. (2.35) and (2.38) have been used in the third step. A factor $\sin(\Delta\psi)/2$ comes from the development of $\cos(\psi - \Delta\psi)\sin(\psi)$, the other term of this development averages to zero as the phases ψ on successive passages are uniformly distributed between 0 and 2π. To express the square of x_s we use again the sampling theorem [Eq. (2.14c)] and write:

$$x_{s,PU}^2 = \beta_{PU}\left(\frac{1}{N_s}\sum\varepsilon_i^{1/2}\cos(\psi_i)\right)^2 \rightarrow \frac{\beta_{PU}}{N_s}\frac{\varepsilon_{rms}}{2} \tag{2.40}$$

Here, another factor $1/2$ comes by averaging $\cos^2(\psi)$ over the betatron phases present in the sample. Collecting terms we have finally:

$$\Delta\varepsilon = \frac{1}{N_s}\left[-g\sqrt{\frac{\beta_{PU}}{\beta_K}}\sin(\Delta\psi)\varepsilon + \frac{g^2}{2}\frac{\beta_{PU}}{\beta_K}\varepsilon_{rms}\right] \tag{2.41}$$

We take a test particle with $\varepsilon = \varepsilon_{rms}$, include undesired and desired mixing through the—by now familiar—factors $(1 - \tilde{M}^{-2})$ and M, and introduce noise by the factor U, determined by an equivalent sample error at the pickup. Further we call $g\sqrt{\beta_{PU}/\beta_K} = g_\beta$, and use Eqs. (2.2) and (2.11) to express N_s by N and $\Delta\varepsilon/\varepsilon$ by $1/\tau$. Finally we determine the cooling rate for the "amplitude" $\sqrt{\varepsilon}$ rather than for ε. Then

$$\frac{1}{\tau} = \frac{1}{2}\frac{W}{N}\left[2g_\beta(1 - \tilde{M}^{-2})\sin(\Delta\psi) - g_\beta^2(M + U)\right] \tag{2.42}$$

Compared to Eq. (2.30) the factors $1/2$ in front and $\sin(\Delta\psi)$ in the coherent term are new. It is clear from the analysis that they are due to the oscillatory nature of the particle motion. On average, the particle reveals only half the information on its amplitude when passing the pickup and the position has to transform to an angular error to be corrected at the kicker. The optimum phase-advance pickup to kicker is

$\pi/2$ (modulo π), as anticipated in Sect. 2.1. In analogy with Eq. (2.31) the cooling rate, Eq. (2.42), has a maximum which is proportional to $[(1 - \tilde{M}^{-2})\sin(\Delta\psi)]^2$. Thus, for imperfect spacing, the inefficiency is given by $\sin(\Delta\psi)^2$.

2.13 Hereward Cooling

In the analysis of Palmer cooling, we assumed that the momentum contribution $x_p = D\Delta p/p$ of the particles dominates at the pickup, whereas for transverse cooling we supposed that the betatron part $x_\beta = \sqrt{\varepsilon\beta}\cos(\psi)$ was most important. The more realistic case, where both x_p and x_β are present, will reveal [12] the mutual heating as well as the fact, that the Palmer system can also be used for transverse cooling. Here we briefly sketch the derivation taking for simplicity $\beta' = 0$ and $D' = 0$ at the kicker.

A question that comes to mind is: How can the acceleration gap used in the Palmer system change the betatron oscillation? The answer is sketched in Fig. 2.13: a particle of nominal momentum oscillates around the central orbit, whereas an off-momentum particle bounces around the off-momentum orbit displaced by $D\Delta p/p$. If at a gap the momentum changes abruptly, the particle continues its oscillation around this displaced orbit with initial conditions $x = x_0 - D\Delta p/p, x' = x_0'$ as its oscillation centre has jumped by $D\Delta p/p$ ($D'(s) = 0$ and $\beta' = 0$ assumed). Thus, the mechanism is similar to changing the oscillation of a pendulum by jumping its point of suspension.

We can now write down the basic equations for Hereward cooling. The RF gap changes the momentum by an amount proportional to the observed position error of the sample. The basic equation for the change of the momentum displacement and the betatron displacement from the instantaneous orbit is

$$\frac{1}{D_K}\Delta x_{p,K} = -g\frac{1}{D_{PU}}\langle x \rangle_{s,PU}$$

$$\frac{1}{D_K}\Delta x_{\beta,K} = g\frac{1}{D_{PU}}\langle x \rangle_{s,PU} \tag{2.43}$$

$$\Delta x'_{p,K} = \Delta x'_{p,PU} = 0$$

In Eq. (2.43) $\langle x \rangle_s = \langle x_p + x_\beta \rangle_s$ is the total pickup signal of a sample containing both the betatron and the momentum contribution. We simplify, again assuming zero derivative of $D(s)$ at the kicker, but we allow for different D at pickup and kicker. To apply our basic procedure, we once more refer the observation at the pickup forward to the kicker and obtain for the momentum part of the first equation (using $x_p/D_{PU} = x_p/D_K$)

$$\Delta x_p = -g\langle x_p \rangle_s$$

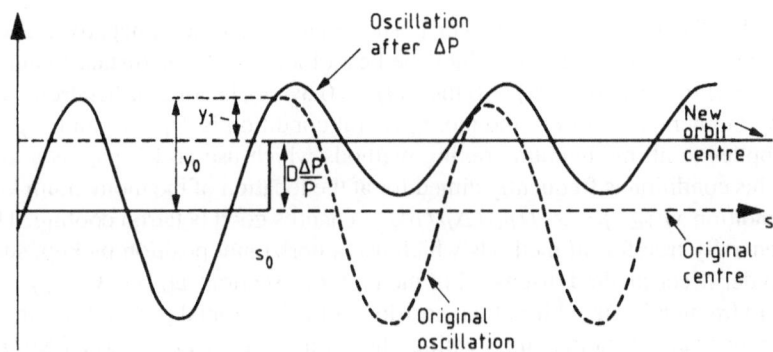

Fig. 2.13 Reduction of betatron oscillation by a momentum jump. At s_0 the momentum is abruptly changed by $\Delta p/p$. The particle continues its betatron oscillation around the new orbit centre, which is displaced by $D(\Delta p/p)$. Initial conditions for this new oscillation are $x_1 = x_0 - D\Delta p/p, x_1' = x_0'$ (change of lattice functions: $D' = 0$ and $\beta \Delta' = 0$ at kicker assumed) (By courtesy of CERN, (©) CERN)

By our standard procedure this leads to the cooling rate for the mean square momentum deviation:

$$\frac{1}{\tau_p} = \frac{W}{N}\left[2g\left(1 - \tilde{M}^{-2}\right) - g^2\left(M + \frac{x_{\beta,rms,PU}^2 + x_{n,PU}^2}{D_{PU}^2(\Delta p/p)_{rms}^2}\right)\right] \qquad (2.44)$$

Here noise (x_n^2) and mixing are included. The term $x_{n,PU}^2/D_{PU}^2(\Delta p/p)_{rms}^2$ is the usual noise to signal ration U. The term $x_{\beta,rms,PU}^2/(D_{PU}^2(\Delta p/p)_{rms}^2)$ shows the momentum heating by the betatron oscillation. In working out Eq. (2.44) we have assumed absence of correlation between betatron and momentum oscillations.

To work out betatron cooling we refer the observation [betatron part of r.h.s. of second Eq. (2.43)] to the kicker, located $\Delta\psi$ downstream, and write the betatron part of Eq. (2.43) as

$$\frac{1}{D_k}\Delta x_{\beta,K} = g\frac{1}{D_{PU}}\sqrt{\frac{\beta_{PU}}{\beta_K}}\langle x_\beta\rangle_{s,K}\cos(\Delta\psi) \qquad (2.45)$$

The uncorrelated momentum part acts as noise. Going once more through the standard procedure we find the cooling rate for the r.m.s. betatron amplitude

$$\frac{1}{\tau_\beta} = \frac{1}{2}\frac{W}{N}\left[-2g_h\left(1 - \tilde{M}^{-2}\right)\cos(\Delta\psi) - g_h^2\left(M + \frac{D_{PU}^2(\Delta p/p)_{rms}^2 + x_{n,PU}^2}{x_{\beta,rms,PU}^2}\right)\right]$$
$$(2.46)$$

Here $g_h = gD_K\sqrt{\beta_{PU}}/(D_{PU}\sqrt{\beta_K})$. We find again the factor $1/2$ due to the oscillatory nature of the betatron motion. The term $D_{PU}^2(\Delta p/p)_{rms}^2/x_{\beta,rms,PU}^2$ shows the heating of the betatron motion by the momentum oscillations. The $\cos(\Delta\psi)$ together with the minus sign in the coherent term indicate that simultaneous cooling is possible if the distance pickup to kicker is such that the cosine is negative. The distance $\Delta\psi = \pi$ (modulo 2π) provides the best efficiency. If the distance cannot be chosen so as to have $\cos(\Delta\psi) < 0$ then $D_K = 0$ avoids heating of betatron oscillations by momentum cooling (the more general condition is $D_K = 0$ and $D'_K = 0$). This applies to all momentum-cooling methods which use an RF gap as a kicker; hence this condition is frequently aimed for at the location of the momentum kicker. The condition $x_{\beta,rms,PU} \gg D_{PU}(\Delta p/p)_{rms}$ ensures good betatron cooling. This is also generally true for all methods which use a horizontal position pickup, such as the van der Meer method discussed in the previous section; $D_{PU} = 0$, $D'_{PU} = 0$ is therefore frequently aimed for at the position of the horizontal pickup for transverse cooling. For Palmer momentum cooling, the inverse $x_{\beta,rms,PU} \ll D_{PU}(\Delta p/p)_{rms}$ is favoured to minimise the 'betatron noise'. For combined horizontal and longitudinal cooling by the Hereward method a compromise ($x_{\beta,rms} \approx x_{p,rms}$ at the pickup) has to be found.

These considerations provide some insight into the problems of 'cross-heating'. We mention two other tricks which have been considered for avoiding horizontal heating by momentum cooling: (i) splitting the momentum kicker into two units spaced by $\Delta\psi = \pi$ or (ii) a horizontal kick proportional to the momentum kick. In the second case the compensator has to be placed at $\Delta\psi = \pi/2$ from the momentum kicker ($\Delta\psi = 0$ if D' is large at the kicker). These methods require that mixing over the distance $\Delta\psi$ is negligible. If the mixing is imperfect, it is (sometimes) also possible to decouple the cross-heating as the longitudinal and transverse signals have different frequencies. This becomes clearer in the frequency-domain analysis.

2.14 Momentum Cooling, Thorndahl's, Kells's and Möhl's Methods

The filter method [21, 22] of Thorndahl and Carron uses the flight time over one revolution to assess the momentum deviation. The method proposed by Kells [23] is based on the flight time between pickup and kicker. Both methods are more naturally analysed in frequency-domain (Appendix A). Here we only give a very simplified introduction. The arrangement for filter cooling is sketched in Fig. 2.14. The filter is a periodic notch filter ('comb') with ideally zero transmission at the harmonics of the revolution frequency in the pass-band.

The device originally used by Thorndahl and Carron is a TEM line of electrical length equal to $1/2$ of the circumference of the storage ring (C), divided by the velocity factor β of the particles (i.e. $\ell = \frac{1}{2}C/\beta$), such that signals travel to the end of the line in half of the nominal revolution time. The line is shorted at the far end ("stub line"); its open end is connected to the cooling path so that—for the synchronous particle ($\delta p = 0$)—the reflected signal from the previous turn annihilates the direct signal. For particles with finite Δp the cancellation is only partial and acceleration or deceleration occurs.

Modern cooling filters (proposed by G. Carron) use a correlator as sketched in Fig. 2.15. These are easier to set up and to control than the "stub line". They are built

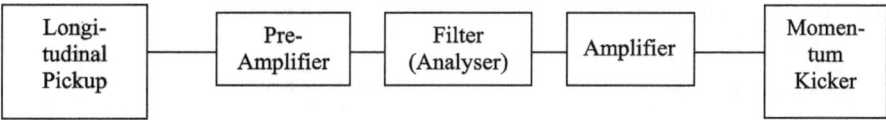

Fig. 2.14 The set-up for filter cooling. The signal from the sum pickup is pre-amplified, and compared in an analyser with the signal from the previous turn. The correction signal is sent, via a final amplifier, to the momentum kicker. A particle of nominal momentum (nominal time between actual and previous passage) is not affected by its own signal. Particles which are too slow (or to fast) are accelerated (or decelerated)

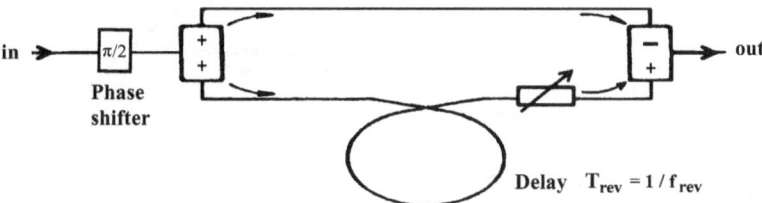

Fig. 2.15 The correlator type notch filter, consisting of a phase shifter, a splitter, a short and a long transmission line and a "subtractor" (combiner with a 180° phase-shift in one branch). For filter cooling, the phase shifter introduces 90 degree shift (equivalent to a differentiation of the signal). The difference in delay between the short and long lines corresponds to the nominal revolution time. The lines, especially the long one, have very small losses. The resistance can be used, to adjust the depth and width of the notches

of high quality,—sometimes even cryogenically cooled—TEM or optical lines. The correlator is preceded by a 90° phase shifter. Then the "cancellation" leads to a transmission characteristic, sketched in Fig. 2.16, that is linear for small momentum deviation.

In fact, as will become clearer in the frequency-domain analysis, one can approximate the pickup pulse of a particle by a cosine like function with a width inversely proportional to the maximum frequency of the pass-band. The differentiator (90° phase shifter) transforms this signal into a sine like pulse at the entrance of the correlator. The outgoing signal is the difference between this pulse and the pulse from the previous turn. The arrangement is tuned such that for a synchronous particle the outgoing signal vanishes (taking into account the proper travelling times). Then for an off-momentum particle it has a delay $\delta T = T \eta \delta p/p$. Here T and η refer to a full revolution. The correlated signal has still to travel to the kicker. It accumulates an additional delay $\delta T_{PK} = T_{PK} \eta_{PK} \delta p/p$. The total delay between an off-momentum particle and its correction signal then amounts to

$$\frac{\delta t}{T_{rev}} = \left\{ \eta + \frac{t_{PK}}{T_{rev}} \eta_{PK} \right\} \frac{\delta p}{p} \equiv \eta_{eff} \frac{\delta p}{p} \tag{2.47}$$

Eq. (2.47) defines the leading term with $\eta_{eff} = \eta + (t_{PK}/T_{rev})\eta_{PK}$ to be used in Fig. 2.16. Another term with $\eta_1 = \eta_{PK} t_{PK}/T_{rev}$ has to be subtracted (see Appendix B). For the example of a regular lattice with $\eta_{PK} = \eta$ and a cooling loop

Fig. 2.16 Correction signal $S \propto \sum_n[\sin\{2\pi n\eta_{eff} \cdot \delta p/p\} - \sin\{2\pi n\eta_1 \cdot \delta p/p\}]$ of a test-particle as a function of its momentum error $n_{max}\eta_{eff}\delta p/p$. A one-octave system ($f_{max} = 2f_{min}$) with an ideal notch filter is assumed. The upper graph is for $\eta_1 = \eta_{eff}/3$, the lower for $\eta_1 = 0$

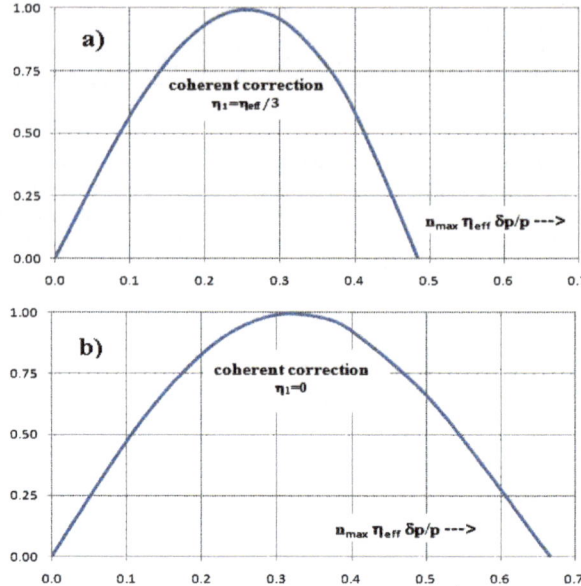

that cuts diagonally through the storage ring one has for instance $\eta_{eff} = 1.5\eta$ and $\eta_1 = \eta_{eff}/3$ as assumed for Fig. 2.16(a). The coherent effect, Fig. 2.16, limits the momentum spread as a function of η and bandwidth (or, more precisely, the upper harmonic $n_{max} = f_{max}/f_{rev}$).

$$(\delta p/p)_{lim} \approx \pm \frac{const}{\eta_{eff} n_{max}} \tag{2.48}$$

One deduces from Fig. 2.16(a), that good linearity is retained, if the constant in Eq. (2.48) has a value const ≤ 0.15, ($\delta p/p \leq \pm 1.5 \times 10^{-3}$). The correction weakens beyond the maximum and becomes unusable for const $> 0.5(\delta p/p > \pm 5 \times 10^{-3})$. Here the $\delta p/p$ values in the brackets are for a system with $\eta_{eff} = 0.05$, $\eta_1 = 0.017$ and $n_{max} = 2000$, corresponding e.g. to a 1–2 GHz system in a 300 m circumference ring (roughly the parameters of the RESR, planned at GSI Darmstadt). Although Figs. 2.16(a) and (b) are strictly only valid for an octave frequency range; the part up to $n_{max}\eta_{eff}\delta p/p \approx 0.5$ depends only weakly on the bandwidth and can be used as a general approximation. If $\eta_1 = \eta_{PK}t_{PK}/T$ is much smaller than $\eta_{eff}/3$ then the initial linearity and the zero crossing extend, to const ≈ 0.2 and ≈ 0.7 in Figs. 2.16(a) and (b) respectively.

The coherent effect thus depends on the momentum deviation $n_{max}\eta_{eff}\delta p/p$ of the particle, and to calculate the cooling of a beam one has to average. Simplifying again by taking a particle with r.m.s.-momentum deviation as typical we can again apply the standard procedure and obtain an equation of the form (2.30) where

$$1 - \tilde{M}^{-2} \approx \cos[0.5\pi \eta_{eff} n_{max}(\Delta p/p)_{rms}]. \tag{2.49}$$

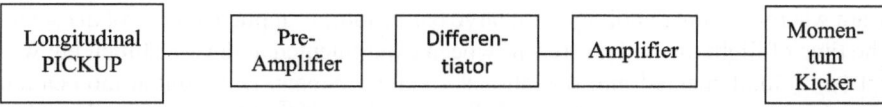

Fig. 2.17 The set-up for time of flight cooling. The signal from the sum pickup is pre-amplified, differentiated and sent, via a final amplifier, to the momentum kicker. A particle of nominal momentum (nominal flight time pickup to kicker) is not affected by its own signal. Particles which are too slow (or to fast) are accelerated (or decelerated)

For the desired mixing factor M ("number of turns for a particle with one standard deviation in momentum to migrate by one sample length $1/(2W)$") we use for the moment again the rough approximation

$$M = \frac{1}{2WT\eta(\Delta p/p)_{rms}} \qquad (2.50)$$

We then obtain an equation similar to Eq. (2.21). However several features specific to the present case have to be mentioned: sum pickups are used and they have a higher sensitivity than the difference devices needed for Palmer cooling. The noise is also reduced by the filter and therefore less important in the centre of the distribution. On the other hand—as discussed above—Eq.(2.48) introduces a rather stringent limit on $\eta f_{max}\Delta p/p$.

This latter restriction is weakened in the time of flight method (Fig. 2.17). Here the pickup pulse of a particle is again transformed to a sine like pulse by a differentiator and then sent directly to the kicker. The delay is adjusted in such a way that a particle of nominal momentum (nominal flight time pickup to kicker) is not affected by its own signal. Particles which are too slow (or to fast) are accelerated (or decelerated). One notes that the time of flight method is equivalent to 'the correlation filter method (Fig. 2.15) without the long signal path'.

We can use Fig. 2.16(b) inserting $\eta_{eff} = (t_{PK}/T_{rev})\eta_{PK}$ to exemplify the coherent signal as a function of $\Delta p/p$. In contrast to Eq. (2.47), the momentum deviation, up to which the coherent signal is approximately linear, is now given by the time of flight and the η-value between pickup and kicker.

$$(\delta p/p)_{lim} \approx \pm \frac{const}{\eta_{PK}f_{max}t_{PK}/T_{rev}} \qquad (2.51)$$

As an example we take again a regular lattice (i.e. $\eta_{PK} = \eta$) and a cooling loop that cuts diagonally through the ring ($t_{PK} = \frac{1}{2}T_{rev}$). Then there is a gain of a factor of 3 in the maximum momentum spread which the system can handle with the same cooling band. As in the filter method sum pickups with their high inherent sensitivity are used. However the electronic noise is—a priory—not reduced, in contrast to filter cooling.

Comparing the different momentum cooling approaches: The filter method has been widely exploited for low intensity beams (e.g. in the AA, LEAR and the AD at CERN). In the antiproton accumulators, it is used for pre-cooling the low intensity beam whereas Palmer cooling (with large bandwidth) is applied to control the stack. The time of flight method has recently been experimented at the F.Z. Jülich but was not (yet?) implemented into operational cooling systems. If the initial momentum spread is large one may think of starting with time of flight and "switch" to filter cooling when $\Delta p/p$ has sufficiently decreased.

A further momentum cooling method (first proposed by the author but not tried out) is based on the difference signals from 2 sum pickups, separated only by a fraction of a turn. This arrangement alleviates the phase distortions caused by the one-turn delay of the filter method (see Fig. 2.16).

Chapter 3
Pickup and Kicker Impedance

3.1 Definitions

A detailed analysis of pickup and kicker structures used for stochastic cooling can be found in the article by Lambertson [24]. Here we only give a summary concentrating mainly on the 'loop couplers'.

For a *sum pickup* (applied for longitudinal cooling) we use the power output from the pick-up divided by the square of the beam current as general definition of the (frequency and geometry dependent) pickup transfer impedance:

$$Z_{P||} = P/I_{beam}^2 \tag{3.1}$$

The special case of a *sum pickup of loop coupler type* (Fig. 3.1) may be visualised as an upper and a lower strip-line, each formed by the electrode and its adjacent ground plane (chamber). Each of the two strip-lines has a characteristic impedance Z_0 the value of which is slightly influenced by the coupling to the other line. Typical values are $Z_0 = 50\ \Omega$.

The signals from the upper and lower electrode are added. They may be regarded as originating from a single source with the characteristic impedance $Z_0/2$. The pickup plates are terminated by Z_0, each, at the downstream end and the signal is coupled out at the upstream end. The units have a typical length $\ell = \frac{1}{4}v_{eff}/f_m$ ("quarter wavelength at mid-band") where f_m is the band middle frequency and $v_{eff} \approx c$ is a velocity given by $1/v_{eff} = \frac{1}{2}(1/v_{line} + 1/v_{particle})$. Usually the pick-ups for cooling consist of many units the output of which is added in combiners.

To obtain the output power P we have to include a geometry factor $g_{||}$ and a frequency dependent factor $F(\omega)$ (see next subsections). We write Eq. (3.1) as

$$Z_{P||} = g_{||} F(\omega) Z_0/2 \tag{3.2}$$

For a beam centred in a sum pickup with $W \gg h$ one has $g_{||} = 1$, and at 'mid-band' also $F(\omega) = 1$ for the loop arrangement (Fig. 3.1). In this situation we have the maximum

$$Z_{P||,m} = Z_0/2 \tag{3.3}$$

D. Möhl, *Stochastic Cooling of Particle Beams*, Lecture Notes in Physics 866, DOI 10.1007/978-3-642-34979-9_3, © Springer-Verlag Berlin Heidelberg 2013

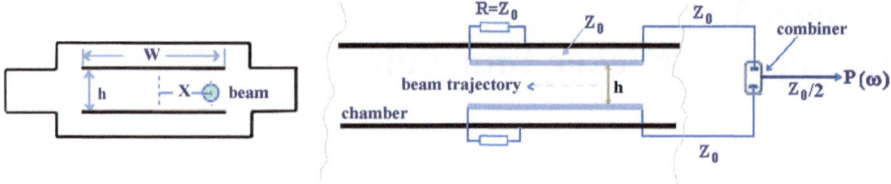

Fig. 3.1 Sum pickup, radial and longitudinal chamber cross-section. For a difference pickup the signals from upper and lower electrode are added in counter-phase

Similar to Eq. (3.1) we define the transfer impedance of a general *sum kicker* by the energy gain $\Delta E = eV_{beam}$ given to the particles upon a passage through the kicker, and the power P_k needed to excite the kicker:

$$Z_{k\|} = V_{beam}^2 / P_k \tag{3.4}$$

The *loop type sum kicker* has a structure similar to the pickup (Fig. 3.1) except that the termination is at the upstream end. The structure can again be regarded as an upper and lower strip-line, each terminated with its characteristic impedance $R = Z_0$. The amplifier (connected to the downstream end) works into a load of impedance $Z_0/2$ and delivers the power $P_k = V_{line}^2/(Z_0/2)$ to the kicker. Particles receive (for $x \approx 0$ and $W \gg h$, Fig. 3.1) a kick $eV_{line}(t)$ at the entrance and $eV_{line}(t + \ell/v_{line} + \ell/v_{beam})$ at the exit of the kicker. These kicks add and at midband where $(1/v_{line} + 1/v_{beam})\omega\ell = \pi/4$, such that $eV_{beam} = 2eV_{line}$. Then the kicker transfer impedance is given by

$$Z_{K\|,m} = 2Z_0 \tag{3.5}$$

We use this maximum impedance (which is $4\times$ the pickup impedance Eq. (3.3)!) to characterise the sum loop kicker. For more general beam position and frequency we have again to include a geometry factor $g_{//}$ and a frequency dependent $F(\omega)$ and write for the *loop type sum kicker*:

$$Z_{K\|} = g_{\|}F(\omega)2Z_0 \tag{3.6}$$

It is shown in the literature [24] that the functions $g_{//}$ and $F(\omega)$ are the same for sum pickup and sum kicker structures ('reciprocity theorem').

We now turn to the *difference mode* (transverse cooling) again referring to the pickup sketched in Fig. 3.1, where now the difference between upper and lower electrode is taken. Hence the output depends on the beam displacement (y in Fig. 3.1). We characterise this by a geometry factor $g_{\perp} \approx \frac{y}{h/2}$ (more general expressions for $g_{\perp}$ will be given in Sect. 3.2 below). We define the pickup impedance by the power output per squared beam current of a thin (pencil) beam passing close to $y = h/2 \rightarrow g_{\perp} = 1$:

$$Z_{P\perp} = P/I_{beam}^2 \tag{3.7}$$

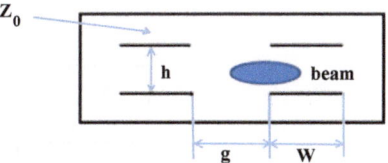

Fig. 3.2 Horizontal difference pickup consisting of two sum pickups. Radial chamber cross-section. The two left plates are on the same potential U and the two right ones on $-U$

For the special case of a *difference pick up of loop type* we have

$$Z_{P\perp} = g_\perp F(\omega) Z_0/2 \tag{3.8}$$

The geometry and frequency factors will be established in the next subsections. We mention here that $F(\omega)$ is the same function for the three cases mentioned (loop sum and difference pickup and loop sum kicker)

As alternative to the arrangement of Fig. 3.1, where the difference between upper and lower electrode is taken, one can use the geometry sketched in Fig. 3.2 as difference pickup. It may be viewed as two sum pickups with a gap in between working in 'push-pull' mode. In this case the electrodes are parallel to the direction of the oscillation to be recorded and ample space in this direction can be made available. For this second type of difference device we write

$$Z_{P==} = g_{==} P/I_{beam}^2 \tag{3.9}$$

For the special case of using the electrodes in loop coupler arrangement

$$Z_{P==} = g_{==} F(\omega) Z_0/4 \tag{3.10}$$

Here Z_0 is the characteristic impedance between each of the 4 plates and the chamber, $F(\omega)$ is the same frequency function used above and $g_{==}$ is another geometry factor (see Sect. 3.2 below)

Finally we turn to *difference kicker*. We use "deflection Voltage"

$$V_{beam} = V_\perp \equiv \beta \Delta p_\perp c/e = \int_s (E_y + \beta c B_x) ds \tag{3.11}$$

and define the impedance

$$Z_{k\perp} = V_\perp^2/P_{k\perp} \tag{3.12}$$

similar to Eq. (3.4) for the longitudinal kicker. Here $P_{k\perp}$ is the power to excite the kicker. The deflection angle at the kicker is

$$\frac{\Delta p_\perp}{p} \equiv \theta = \frac{e V_\perp}{\beta p c} \tag{3.13}$$

Fig. 3.3 Sketch of the
structure to construct
geometry factors

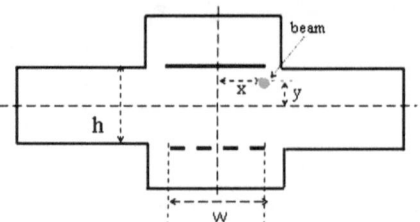

The difference *loop kicker* has a structure identical to the pickup except that once again the termination is at the upstream end. We use a relation between difference pickup and kicker which we obtain from considerations given in Ref. [24] (Eq. (4.19) of Goldberg and Lambertson, valid if $\partial Z_{P\perp}/\partial x \ll \partial Z_{P\perp}/\partial y$)

$$Z_{k\perp} = -2i\frac{v_b}{\omega}\frac{\partial Z_{P\perp}}{\partial y} \tag{3.14}$$

The factor $2i\frac{v_l}{\omega}$ in Eq. (3.14) changes the frequency response (next subsection), the derivative modifies the geometry factor. We write for the *transverse loop kicker*

$$Z_{K\perp} = 2Z_0 g_{K\perp} F_{K\perp}(\omega)$$

$$\text{with } g_{K\perp} = \frac{\partial g_\perp}{\partial y}, \; F_{K\perp}(\omega) = -2i\frac{v_b}{\omega}F(\omega) \tag{3.15}$$

For the second type of transverse kicker (Fig. 3.2) we can establish similar relations.

3.2 Geometry Factors

Following R. Shafer [25] we construct the different sensitivity factors from the module sketched in Fig. 3.3, where only the upper electrode is excited. Its sensitivity factor is

$$e(x, y) = \frac{1}{2\pi}\left\{\arctan\left(\frac{\sinh(\frac{\pi}{h}(x + \frac{w}{2}))}{\cosh(\frac{\pi y}{h})}\right) - \arctan\left(\frac{\sinh(\frac{\pi}{h}(x - \frac{w}{2}))}{\cosh(\frac{\pi y}{h})}\right)\right\}$$

$$+ \frac{1}{2\pi}\left\{\arctan\left(\tan\left(\frac{\pi y}{h}\right)\tanh\left(\frac{\pi}{h}\left(x + \frac{w}{2}\right)\right)\right)\right.$$

$$\left. - \arctan\left(\tan\left(\frac{\pi y}{h}\right)\tanh\left(\frac{\pi}{h}\left(x - \frac{w}{2}\right)\right)\right)\right\} \tag{3.16}$$

From Eq. (3.16) we construct the expressions as indicted in Table 3.1. Examples of different sensitivity factors are displayed in Figs. 3.4, 3.5, 3.6 and 3.7.

Table 3.1 Sensitivity of different pickup structures as constructed from $e(x, y)$, Eq. (3.16)

Structure	Sensitivity	Pickup impedance	Kicker impedance
Combination of formula for loop couplers			
Sum	$g_{//} = e(x, y) + e(x, -y)$	$Z_{P//} = \frac{1}{2} Z_0 g_{//} / F(\omega)$	$Z_{K//} \approx 2 Z_0 g_{//} / F(\omega)$
Difference 1(*)	$g_\perp = e(x, y) - e(x, -y)$	$Z_{P\perp} = \frac{1}{2} Z_0 g_\perp F(\omega)$	$Z_{K\perp} \approx 2 Z_0 \frac{\partial g_\perp}{\partial y} F_{K\perp}(\omega)$
Difference 2	$g_{==} = [e(x + w + g, y) + e(x, -y)] - [e(x + w + g, y) + e(x + w + g, -y)]$	$Z_{P==} = \frac{1}{4} Z_0 g_{==} F(\omega)$	$Z_{K==} \approx Z_0 \frac{\partial g_{==}}{\partial x} F_{K\perp}(\omega)$
Detailed formulae for sensitivity			
Sum	$g_{//} = \frac{1}{\pi} \left\{ \arctan\left[\frac{\sinh\frac{\pi}{h}(x + \frac{w}{2})}{\cosh(\frac{\pi}{h} y)}\right] - \arctan\left[\frac{\sinh\frac{\pi}{h}(x - \frac{w}{2})}{\cosh(\frac{\pi}{h} y)}\right] \right\}$		
Difference 1(*)	$g_\perp = \frac{1}{\pi} \left\{ \arctan\left[\tan(\frac{\pi}{h} y) \tanh\left(\frac{\pi}{h}(x + \frac{w}{2})\right)\right] - \arctan\left[\tan(\frac{\pi}{h} y) \tanh\left(\frac{\pi}{h}(x - \frac{w}{2})\right)\right] \right\}$		
Difference 2	$g_{==} = \frac{1}{\pi} \left\{ \arctan\left[\frac{\sinh(\frac{\pi}{h}(\Delta x + \frac{g}{2} + w))}{\cosh(\frac{\pi}{h} y)}\right] - \arctan\left[\frac{\sinh(\frac{\pi}{h}(\Delta x + \frac{g}{2}))}{\cosh(\frac{\pi}{h} y)}\right] \right\} - \frac{1}{\pi} \left\{ \arctan\left[\frac{\sinh(\frac{\pi}{h}(-\Delta x + \frac{g}{2} + w))}{\cosh(\frac{\pi}{h} y)}\right] - \arctan\left[\frac{\sinh(\frac{\pi}{h}(-\Delta x + \frac{g}{2}))}{\cosh(\frac{\pi}{h} y)}\right] \right\}$		
Δx = horizontal distance form center of pickup			

(*) Note that Lambertson, Ref. [24], defines $g_{\perp lambertson} = g_\perp h/(2y)$ (Fig. 3.1)

Fig. 3.4 Geometrical sensitivity ($g_{//}$) of a sum pickup (Fig. 3.1) as function of the horizontal distance (x) from the pickup centre for different width (W) of the pickup plates. The beam is vertically centred ($y = 0$). In normal applications the beam is near the centre of the pickup, where for a ratio width to height (w/h) larger 2, one has $g_{//}$ close to 1 (*upper graph*). For stacking applications, part of the beam is at large distance where the sensitivity falls off exponentially (*lower graph*)

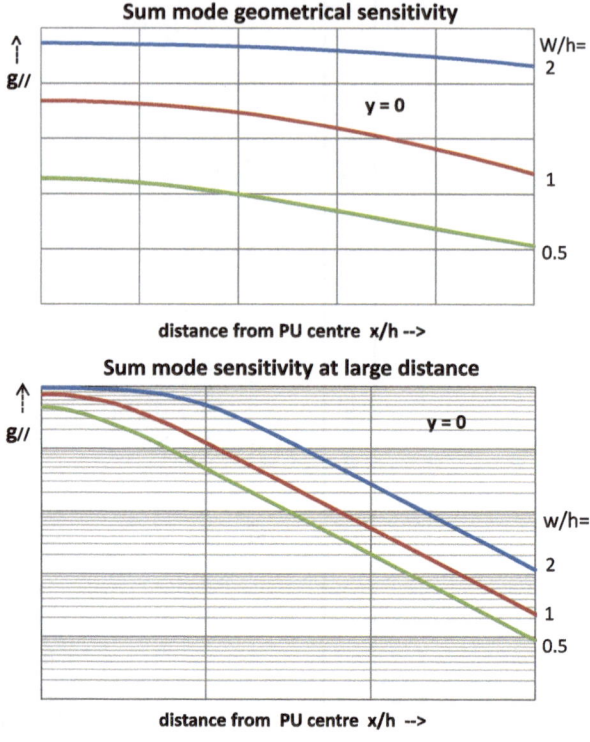

Fig. 3.5 Sensitivity function of a difference pickup as sketched in Fig. 3.1. The response is shown as a function of the vertical distance from the medium plane. Three different widths (w) of the pickup-plates are assumed. The beam is horizontally centred ($x = 0$)

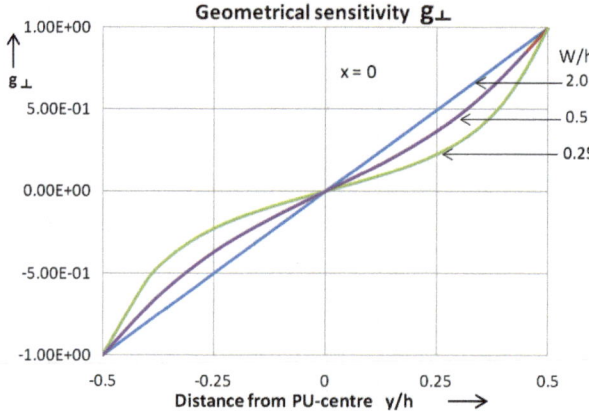

3.3 Frequency Response

The frequency response depends on the way the lines, formed by plate and chamber, are terminated at the two ends. Here we will assume loop couplers, where the impedance of the input and output connection matches the characteristic impedance of the lines.

Fig. 3.6 Sensitivity function ($g_{==}$) of a difference pickup as sketched in Fig. 3.2. The response is shown as a function of the horizontal distance from the centre. Three different widths of the pickup-plates are assumed. The gap "g" (see Fig. 3.2) is constant ($g/h = 0.5$) and the beam is vertically centred ($y = 0$)

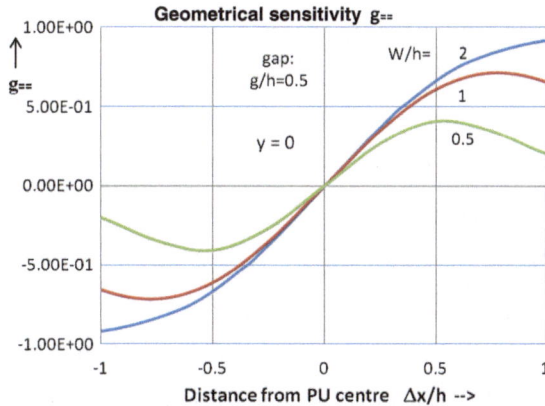

For the sum mode, the interaction takes place at the entrance and exit of the pickup. Away from these ends the fields between the plates and their adjacent ground plane (chamber walls) are purely transverse (TEM-wave) and practically zero in the inner space at the particle orbits. The TEM-waves propagate at the velocity v_l of the line. This velocity would be the velocity of light for smooth two-dimensional conductors, but may be reduced by the presence of magnetic or dielectric media or by longitudinal variations of the cross-section of the conductors. To establish the pickup response we assume a beam current $I_n e^{i\omega t}$. It induces a forward travelling voltage wave $Z_0 I_n e^{i\omega t}$ at the entrance and a backward wave $-Z_0 I_n e^{i\omega(t+l/v_b)}$ at the exit of the pickup (of length ℓ). To add these voltages (referred to the entrance), one has to take the velocity (v_l) at which the backwards wave travels to the entrance into account. One obtains

$$V_n = \frac{1}{2} Z_0 e^{i\omega t} \left(1 - e^{i\omega(l/v_b + l/v_l)}\right)$$

This suggest the frequency response, $F(\omega)$, Eq. (3.2) for the sum mode

$$F(\omega) = \frac{1}{2}\left(1 - e^{i\omega(l/v_b + l/v_l)}\right) \tag{3.17}$$

Eq. (3.17) may also be written as

$$F(\omega) = e^{i(\pi/2 - \Theta)} \sin(\Theta) \quad \text{with } \Theta = \frac{1}{2}\omega(l/v_b + l/v_l) \tag{3.18}$$

This shows the $\sin(\Theta)$ amplitude response and the ($\frac{\pi}{2} - \Theta$) phase response (Fig. 3.7) of the pickup. If both beam and line velocity equal c, then $F_{//}$ is maximum and real for $l = \lambda/4$. For this reason, the electrode arrangement is often called "quarter-wave loop coupler".

Equations (3.17) and (3.18) so far are for a *sum pickup*. However it is shown in Ref. [24] that the same complex function $F(\omega)$ equally applies for the response of the *difference pickup* and also for the *sum kicker* (powered at the upstream end). For

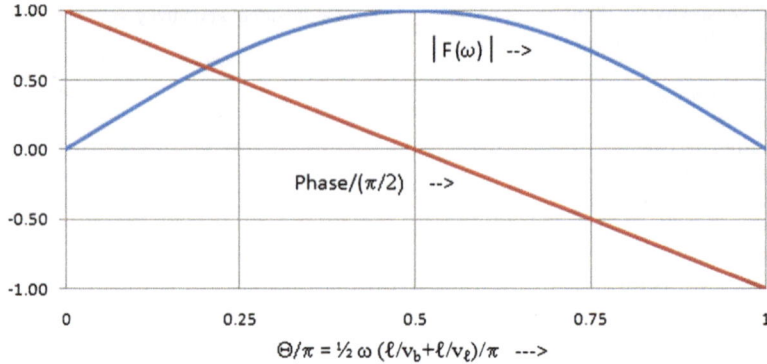

Fig. 3.7 Amplitude and phase response of sum and difference pickup and sum kicker

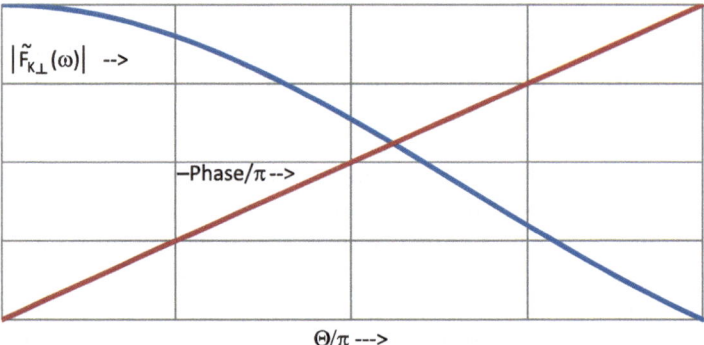

Fig. 3.8 Amplitude and (negative value of) phase response of a transverse loop kicker

the pickup, this is understandable as both the upper and the lower electrode respond in a similar manner to sum or difference mode. They mainly differ by the geometry factors.

To establish the response of the *transverse kicker*, we use Eqs. (3.15) and (3.18):

$$F_{K\perp} = -2i\frac{v_b}{\omega}e^{i(\pi/2-\Theta)}\sin\Theta = 2\frac{v_b}{\omega}e^{-i\Theta}\sin\Theta$$

Remembering the definition of Θ (under Eq. (3.18)) this may also be written as

$$F_{K\perp} = l(1 + v_b/v_l)e^{-i\Theta}\frac{\sin\Theta}{\Theta} \equiv l(1 + v_b/v_l)\cdot\tilde{F}_{K\perp} \qquad (3.19)$$

The factor $l(1 + v_b/v_l)$ may be regarded as an additional geometry constant. The factor $\tilde{F}_{K\perp} = e^{-i\Theta}(\sin\Theta)/\Theta$ describes the frequency response of the difference kicker (Fig. 3.8). It has the maximum at low frequency.

A response function $F_{k\equiv}$ similar to Eq. (3.18) also holds for the second type of transverse kicker. Collecting results we have final expressions for the transfer impedance listed in Table 3.2.

Table 3.2 Transfer impedances (per pickup or kicker unit) of different loop coupler arrangements when the impedance of the signal transfer system is $Z_0/2$. Equations for the geometry factors are given in Table 3.1. The transfer impedances for the difference kickers which are only given for the range where $g_\perp$ or $g_{==}$ vary linearly with distance y or Δx respectively, see Figs. 3.5 and 3.6

Sum pickup	$Z_{P//} = g_{//}\frac{1}{2}Z_0 e^{i(\pi/2-\Theta)}\sin\Theta$
Sum kicker	$Z_{K//} = g_{//}2Z_0 e^{i(\pi/2-\Theta)}\sin\Theta$
Difference pickup 1	$Z_{P\perp} = g_\perp\frac{1}{2}Z_0 e^{i(\pi/2-\Theta)}\sin\Theta$
Difference kicker 1	$Z_{K\perp} = (\partial g_\perp/\partial y)Z_0 l(1+v_b/v_l)e^{-i\Theta}(\sin\Theta)/\Theta$
Difference pickup 2	$Z_{P==} = g_{==}\frac{1}{4}Z_0 e^{i(\pi/2-\Theta)}\sin\Theta$
Difference kicker 2	$Z_{K==} = (\partial g_{==}/\partial x)\frac{1}{2}Z_0 l(1+v_b/v_l)e^{-i\Theta}(\sin\Theta)/\Theta$

$\Theta = \frac{1}{2}\omega(l/v_b + l/v_l)$ with l: length of PU or kicker, v_b, v_l: beam and wave velocity

3.4 Matching to the Cooling Loop

When the impedance of the signal transmission system is matched to the pickup output and the kicker input then the transfer is just determined by Z_P and Z_K as previously defined. If however the input impedance R_p of the network is different from $Z_0/2$ then the transfer impedance $Z_{P//} = $ (system input voltage / beam current) is changed. It becomes

$$Z_{P//} = \frac{1}{2}\sqrt{2R_p Z_0}\,g_{//}F(\omega) \tag{3.20}$$

The situation may be visualised by a matching transformer inserted between the pickup and the subsequent network.

For the kicker, the generalised transfer impedance for a kicker input R_k becomes

$$Z_{K//} = 2\sqrt{\frac{R_K Z_0}{2}}\,g_{//}F_k(\omega) \tag{3.21}$$

It is often more convenient to work with the kicker voltage transfer function $eK = $ (Beam energy change / system out voltage):

$$K_{K//} = Z_{K//}/R_K = 2\sqrt{\frac{Z_0}{2R_K}}\,g_{//}F_k(\omega) \tag{3.22}$$

Obviously $Z_{P//}$, $Z_{K//}$ and $K_{K//}$ are obtained from our previous relations by simple substitutions:

- to obtain $Z_{P,K//}$ we replace $Z_0 \rightarrow \sqrt{2R_{P,K}Z_0}$ in the corresponding expression of Table 3.2;
- to establish the generalised kicker voltage transfer-function $K_{//}$ we replace $Z_0 \rightarrow \sqrt{Z_0/(2R_K)}$ in the expression for $Z_{K//}$.

Table 3.3 Values (per pickup or kicker unit) for the generalised pickup transfer impedance and the kicker impedance and voltage transfer functions. The pickup works into the cooling loop with input impedance R_P, the kicker is fed from the output with impedance R_K. For the geometry factors approximations, valid for a centred beam, are taken

Sum pickup	$Z_{P//} = \frac{2}{\pi} \arctan(\sinh \frac{\pi w}{2h}) \cdot \frac{1}{2}\sqrt{2R_P Z_0} e^{i(\pi/2-\Theta)} \sin\Theta$
Sum kicker	$Z_{K//} = \frac{2}{\pi} \arctan(\sinh \frac{\pi w}{2h}) \cdot 2\sqrt{Z_0 R_K/(2)} e^{i(\pi/2-\Theta)} \sin\Theta$
	$K_{K//} = \frac{2}{\pi} \arctan(\sinh \frac{\pi w}{2h}) \cdot 2\sqrt{Z_0/(2R_K)} e^{i(\pi/2-\Theta)} \sin\Theta$
Difference pickup 1	$Z_{P\perp} = \frac{2y}{h} \tanh(\frac{\pi w}{2h}) \cdot \frac{1}{2}\sqrt{2R_P Z_0} e^{i(\pi/2-\Theta)} \sin\Theta$
	$Z'_{P\perp} = \frac{2}{h} \tanh(\frac{\pi w}{2h}) \cdot \frac{1}{2}\sqrt{2R_P Z_0} e^{i(\pi/2-\Theta)} \sin\Theta$
Difference kicker 1	$Z_{K\perp} = \frac{2}{h} \tanh(\frac{\pi w}{2h}) \cdot \sqrt{Z_0 R_K/(2)}\ell(1+v_b/v_l)e^{-i\Theta}(\sin\Theta)/\Theta$
	$K_{K\perp} = \frac{2}{h} \tanh(\frac{\pi w}{2h}) \cdot \sqrt{Z_0/(2R_K)}\ell(1+v_b/v_l)e^{-i\Theta}(\sin\Theta)/\Theta$
Difference pickup 2	$Z_{P\equiv} = \frac{2\Delta x}{h}\left\{\cosh(\frac{\pi g}{2h})\right\}^{-1} \frac{1}{4}\sqrt{2R_P Z_0} e^{i(\pi/2-\Theta)} \sin\Theta$
	$Z'_{P\equiv} = \frac{2}{h}\left\{\cosh(\frac{\pi g}{2h})\right\}^{-1} \frac{1}{4}\sqrt{2R_P Z_0} e^{i(\pi/2-\Theta)} \sin\Theta$
Difference kicker 2	$Z_{K\equiv} = (\frac{2}{h})\left\{\cosh(\frac{\pi g}{2h})\right\}^{-1} \frac{1}{2}\sqrt{Z_0 R_K/(2)}\ell(1+v_b/v_l)e^{-i\Theta}(\sin\Theta)/\Theta$
	$K_{K\equiv} = (\frac{2}{h})\left\{\cosh(\frac{\pi g}{2h})\right\}^{-1} \frac{1}{2}\sqrt{Z_0/(2R_K)}l(1+v_b/v_l)e^{-i\Theta}(\sin\Theta)/\Theta$

$\Theta = \frac{1}{2}\omega(l/v_b + l/v_l)$ with l: length of PU or kicker, v_b, v_l: beam and wave velocity

In a similar fashion we obtain the generalised *transverse* pickup impedance and kicker voltage transfer-function using analogue substitutions. Results for the different cases are listed in Table 3.3, where approximate values for the geometry factors are used. Also included is the expression of $Z'_{P\perp} = Z_{P\perp}/y$ frequently used later.

3.5 Signal Combination

The signals from several loop couplers can be added using combiners at the pickup output and splitters at the kicker input (with delays matching the particle travelling time). In this case the power adds and the resulting total pickup impedances and voltage signal are

$$Z_{PU} = n_P Z_{Pi}, \qquad V_{PU} = \sqrt{n_P} V_{Pi} \qquad (3.23)$$

with $Z_{Pi} = Z_{P//}$ or $Z_{P\perp}$, the impedances of the unit sections of sum or difference pickups. The number (n_P) of units includes a (complex) factor smaller than 1 due to the inefficiency of combining. In a similar fashion one has for the kicker

$$Z_{KK} = n_K Z_{Ki}, \qquad K_{KK} = \sqrt{n_K} K_{Ki}, \qquad V_{KK} = \sqrt{n_K} V_{Ki} \qquad (3.24)$$

with Z_{Ki}, K_{Ki}, V_{Ki} the quantities corresponding to the kicker unit sections.

If the beam velocity is sufficiently lower than c, one can combine the units in "travelling wave fashion" i.e. in series, replacing the terminating pickup impedance (Fig. 3.1) by the input of the subsequent section. In a similar way a travelling wave kicker is obtained by connecting the units in series. Ideally the voltage adds linearly in this case and the power in square with the number of units. However it is difficult to combine a large number of units in this way, especially if large bandwidth and/or a range of beam velocities is required.

The development of "slotted pickup and kicker structures" has undergone a rapid development [26–34] ever since the proposal of the Faltin structure [35]. The latter may be viewed as a rectangular TEM guide tightly coupled to the beam through slots in the chamber. Compared to loops: for a given beam velocity this and other slotted structures have larger impedance and a behaviour closer to a travelling wave structure. The geometry and frequency factors of complicated structures (and hence the total impedances Z_{PU}, Z_{KK} etc.) are nowadays computed [29, 30] using "Maxwell's equation solvers" like "HFFS" or "MAFIA". Whereas the geometry and frequency factors depend on the nature of the structure, the factor 4 between kicker and pickup remains unchanged.

Horizontal, vertical and sum signals for virtually all types of pickups/kickers may be detected/created on structures located at the same place or at different azimuthally locations, depending on available space and the geometry factors obtainable.

Chapter 4
Frequency-Domain Picture

4.1 Schottky Noise Signals from a Coasting Beam

We start with the 'Schottky' signals [12, 36–39] induced by a beam coasting in the storage ring: The current, detected by a *longitudinal (sum)* pickup, due to a single particle going around at constant revolution frequency can be represented as a sum of a short pulse at each traversal which can be Fourier analysed into harmonics of the revolution frequency $\omega_{rev} = 2\pi f_{rev}$:

$$I_s(t) = e \sum_{k=-\infty}^{\infty} \delta(t - t_p + kT_{rev}) = ef_{rev} \sum_{m=-\infty}^{\infty} e^{im\omega_{rev}(t-t_p)} \quad \text{or equivalently}$$

$$I_s(t) = ef_{rev} + \sum_{n=1}^{\infty} 2ef_{rev} \cos[n\omega_{rev}(t - t_p)] \tag{4.1}$$

The 'passage time' t_p is given by the initial conditions, namely the instant at which the particle was injected. Equation (4.1) can also be written in terms of cosine and sine components:

$$I_s(t) = ef_{rev} + \sum_{n=1}^{\infty} a_n \cos(n\omega_{rev}t) + b_n \sin(n\omega_{rev}t) \tag{4.2}$$

with

$$a_n = 2ef_{rev} \cos(n\omega_{rev}t_p), \qquad b_n = 2ef_{rev} \sin(n\omega_{rev}t_p)$$

Next, assume a particle with a slightly different frequency $\omega_{rev} = \omega_{rev1}$. It will have a similar time and frequency spectrum with lines at kT_{rev1} and $n\omega_1$ respectively (Fig. 4.1). Hence, for a beam with a distribution of revolution frequencies between $\omega_1 = \omega_{rev} - \Delta\omega_{rev}/2$ and $\omega_2 = \omega_{rev} + \Delta\omega_{rev}/2$, one expects a spectrum with bands around each harmonic $n\omega_{rev}$ (Fig. 4.2). The mth band extends from $m(\omega_{rev} - \Delta\omega_{rev}/2)$ to $m(\omega_{rev} + \Delta\omega_{rev}/2)$. Its width is therefore $m\Delta\omega_{rev}$. This

D. Möhl, *Stochastic Cooling of Particle Beams*, Lecture Notes in Physics 866,
DOI 10.1007/978-3-642-34979-9_4, © Springer-Verlag Berlin Heidelberg 2013

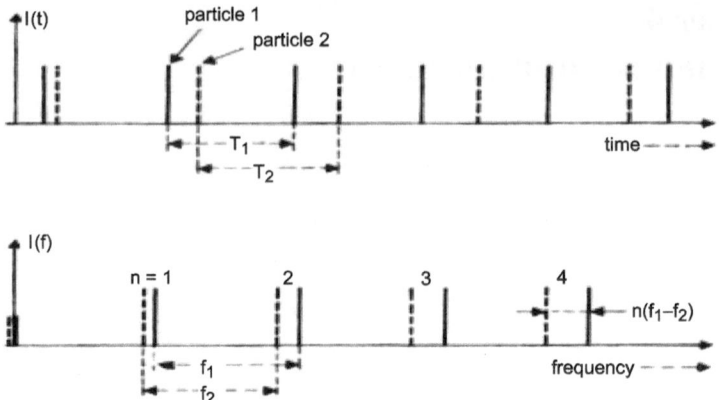

Fig. 4.1 Time and frequency spectrum for a particle of frequency $f_2 = 1/T_2$ and a second particle with $f_1 = f_2 + \Delta f (T_2 > T_1)$ (By courtesy of CERN, (©) CERN)

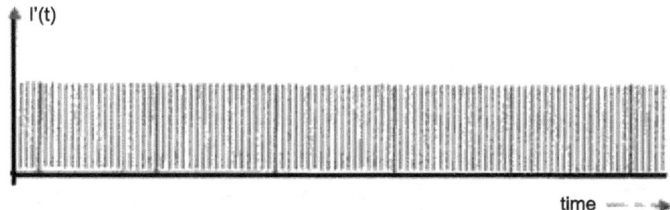

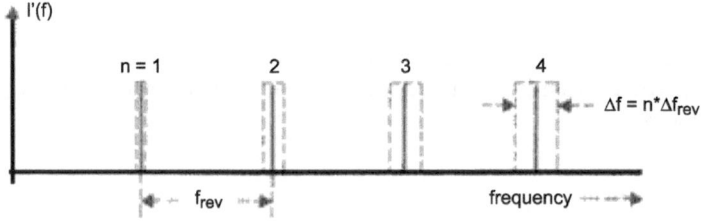

Fig. 4.2 Time and frequency spectrum of a group of particles with a distribution of revolution frequencies $f_{rev} \pm \Delta f/2$. The nth band of the frequency spectrum has a width $n \cdot \Delta f$. The height of the bands is arbitrary at this stage (By courtesy of CERN, (©) CERN)

picture introduces the frequency content of a beam. To calculate the strength of these bands, we have to take the initial phases of the particles ($\omega_{rev} t_p$) into account.

Let us assume a distribution of initial times t_p [Eq. (4.1)] between 0 and the revolution time $T = T_{rev}$, corresponding to an unbunched beam. In the time-domain the pickup signal is then a dense series of pulses which looks like a direct current with,

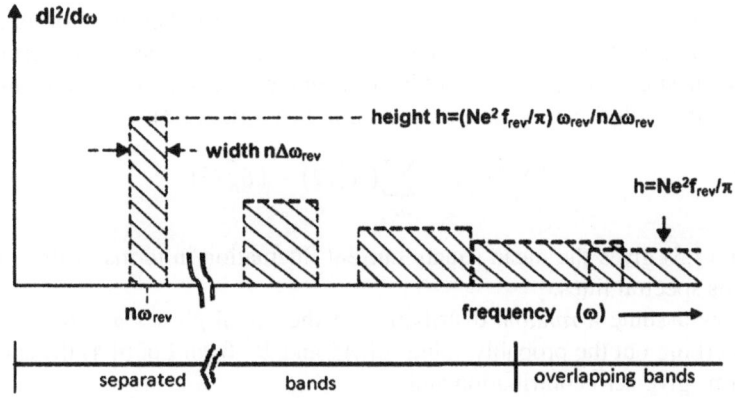

Fig. 4.3 Schottky noise bands (rectangular approximation) of a coasting beam observed on a longitudinal pickup. The band at $n\omega_{rev}$ has a width $n\Delta\omega_{rev}$ and an effective height $(1/\pi)Ne^2 f_{rev} \cdot \omega_{rev}/(n\Delta\omega_{rev})$. At high harmonics ($n > \omega_{rev}/\Delta\omega_{rev}$) bands overlap and the height (spectral density) of the noise power is $(1/\pi)Ne^2 f_{rev}$, as for classical shot noise (By courtesy of CERN, (©) CERN)

superimposed on it, a high-frequency current with a wavelength spectrum given by the spacing of the particles. For *uniformly* spaced particles in typical beams with 10^5 to 10^{10} particles per metre the a.c. component is in the 10^{13}–10^{18} Hz region and is not resolved by the pickup. If, on the contrary, initial phases are distributed in a *random* manner, as they are for instance in a beam emerging from a cathode, then the statistical fluctuation of the number of particles per time interval Δt leads to a noise-like current (Figs. 4.2 and 4.3) with frequency components in the bands discerned above. We shall now try to estimate the density of the noise in these 'Schottky bands' of a coasting beam.

As a first step let us single out a group of N' particles all having the same revolution frequency $f = \omega/2\pi$. The contributions Eq. (4.1) of these particles can be added up to give the current $I'(t)$ of the subgroup:

$$I'(t) = N'e f_{rev} + \sum_{n=1}^{\infty} A_n \cos(n\omega t) + B_n \sin(\omega t) \qquad (4.3)$$

where each of the Fourier coefficients is the sum of the corresponding N' single-particle contributions (with ωt_k the initial phase of particle k):

$$A_n = \sum_{k=1}^{N'} a_{nk} = \sum_{k=1}^{N'} 2e f_{rev} \cos(\omega t_k), \qquad B_n = \sum_{k=1}^{N'} 2e f_{rev} \sin(\omega t_k) \qquad (4.4)$$

In general, we only know the statistical properties of the distribution of initial phases. Our aim is then to derive the corresponding 'probable' values of the Fourier coefficients in order to determine the statistical properties of the current (4.3). We are especially interested in the average current I'_{av} and the mean squared fluctuation

$(\Delta I')^2 = (I'^2 - I'^2_{av})$. This fluctuating noise current—or rather those of its components that fall into the bandwidth of the system—lead to the heating of a test particle during stochastic cooling. We can define the average of $(\Delta I')^2$ over sufficiently long time, say the revolution time or multiples of it. From Eq. (4.3):

$$\langle (\Delta I')^2 \rangle_T = \sum_n (A_n^2/2) + (B_n^2/2) \tag{4.5}$$

Equation (4.5) gives the mean square current fluctuation in terms of the contributions of its spectral lines.

We now assume a *random* distribution of the initial phases $\omega_{rev} t_k \equiv \psi_k$, such that in working out the probably values of A_n^2 and B_n^2 from Eq. (4.4) the sums over cross-terms give zero contribution and

$$A_n^2 = (2ef_{rev})^2 \left(\sum_{k=1}^{N'} \cos n\psi_k \right)^2 = (2ef_{rev})^2 \sum_{k=1}^{N'} (\cos n\psi_k)^2 = (2ef_{rev})^2 N'/2$$

$$B_n^2 = (2ef_{rev})^2 \left(\sum_{k=1}^{N'} (\sin n\psi_k) \right)^2 = (2ef_{rev})^2 N'/2 \tag{4.6}$$

Then each line contributes, according to Eq. (4.5)

$$\langle (\Delta I')^2 \rangle_n = 2(ef_{rev})^2 N' \tag{4.7}$$

Equation (4.7) specifies the intensity of the nth spectral line of the mean-squared noise current for a group of N' mono-energetic (equal ω_{rev}) particles with random initial phases. In reality we are faced with a distribution of revolution frequencies between $\omega_{rev} \pm \Delta\omega_{rev}/2$. Clearly, then, the different components contribute current in the range $n(\omega_{rev} \pm \Delta\omega_{rev}/2)$ around each harmonic n. We thus have a band spectrum, the so-called 'Schottky noise spectrum' (or 'Schottky band spectrum') of a coasting beam. For a subgroup with a very narrow range $d\omega_{rev}$ of frequencies we can apply Eq. (4.7). Interpret $N'd\omega_{rev} = (dN/d\omega_{rev})d\omega_{rev}$ as the fraction of particles with a revolution frequency in the range $\omega_{rev} \pm d\omega_{rev}/2$ and call their contribution to the nth band $d(I)_n^2$. Then from Eq. (4.7)

$$d(I)_n^2 = 2(ef_{rev})^2 (dN/d\omega_{rev})d\omega_{rev}$$

$$d(I)_n^2/d(\omega_{rev}) = 2(ef_{rev})^2 \cdot dN/d(\omega_{rev}) \tag{4.8}$$

$$d(I)_n^2/d\omega_n = 2(ef_{rev})^2 \cdot dN/d(n\omega_{rev}) = 2(ef_{rev})^2 \cdot dN/d\omega_n$$

This specifies the spectral power density of the noise in band n.

Integrating Eq. (4.8) over a band (ω_{rev} going from $\omega_{rev0} - \Delta\omega_{rev}/2$ to $\omega_{rev0} + \Delta\omega_{rev}/2$) and assuming separated bands ($n\Delta\omega_{rev} \ll \omega_{rev}$) we have for the total noise per band,

$$(I)_n^2 = 2(ef_{rev})^2 N \tag{4.9}$$

In words: the area of each Schottky noise band (in a $d(I)_n^2/d\omega$ vs. ω-diagram) is constant. Since the nth band has a width $\Delta\omega = n\Delta\omega_{rev}$ the spectral density decreases with $1/n$. This is also obvious from Eq. (4.8) as we now regard the band on its proper frequency scale $\omega_n = n\omega_{rev}$, such that $dI^2/d\omega = (1/n)dI^2/d\omega_{rev}$). We thus have the situation sketched in Fig. 4.3 where the width of the bands increases and the height decreases linearly with n.

Frequently bands are approximated by a rectangle of width $n\Delta\omega_{rev}$. In this approximation the spectral power density of the noise in band n becomes

$$\frac{d(I)_n^2}{d\omega_n} = \frac{2(ef_{rev})^2 \cdot N}{\omega_{rev}} \cdot \frac{\omega_{rev}}{n\Delta\omega_{rev}} \quad \text{for } n < \omega_{rev}/\Delta\omega_{rev} \qquad (4.10)$$

At high frequency bands overlap and the density is, as for classical shot noise:

$$\frac{d(I)_n^2}{d\omega_{rev}} = \frac{2(ef_{rev})^2 N}{\omega_{rev}} \quad \text{for } n > \omega_{rev}/\Delta\omega_{rev} \qquad (4.11)$$

A practical example of a longitudinal 'Schottky scan' is given by Fig. 1.2 (Chap. 1).

So far we have analysed the Schottky signals obtained by a sum pickup which records the longitudinal motion. For transverse cooling we need the corresponding signals of the position sensitive difference pickup to which we now turn: The output signal can be written $V_s = Z_{PU\perp}I_s$ where $Z_{PU\perp} = Z'_{PU\perp} \cdot x$ is the transfer impedance, introduced in Chap. 3 (Eq. (3.23) and Table 3.3), I_s the longitudinal current Eq. (4.1) and x the displacement of the particle. For a particle performing betatron oscillations:

$$x = \tilde{x}e^{i(Q\omega_{rev}t+\mu_0)}$$

Here μ_0 is the initial betatron phase and $\tilde{x}$ the amplitude. The induced signal is therefore

$$V_s = \tilde{x}ef_{rev} \sum_{m=-\infty}^{\infty} Z'_{PU\perp}(\omega) \cdot e^{i(Q+m)\omega_{rev}t - i(m\omega_{rev}t_p + \mu_0)} \qquad (4.12)$$

Writing Eq. (4.12) with positive m to reveal clearer the $Q \pm m$ bands:

$$V_s = \tilde{x}ef_{rev} \sum_{m=0}^{\infty} Z'_{PU\perp}(\omega) \cdot e^{i(Q\pm m)\omega_{rev}t - i(\pm m\omega_{rev}t_p + \mu_0)} \qquad (4.13)$$

Looking at the exponent we find again that the particle induces signals at the sidebands

$$\omega = \omega_m = \left(m \pm \|Q\| \pm q\right)\omega_{rev} = (n \pm q)\omega_{rev} \qquad (4.14)$$

Here $\|Q\|$ is the integer part and q the non-integer part of the betatron tune; m, $\|Q\|$ and n are all integers.

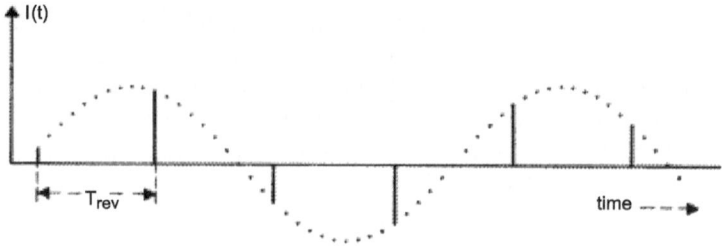

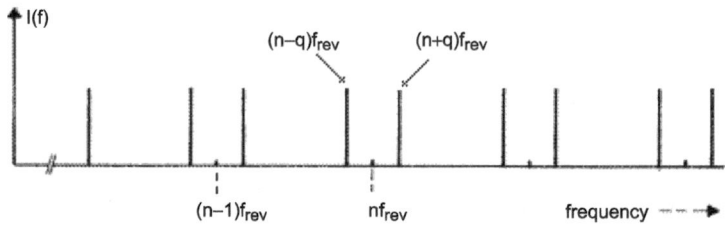

Fig. 4.4 Time and frequency-domain signal of a particle performing betatron oscillations. A position sensitive pickup records a short pulse at each traversal modulated in amplitude by the oscillation. The frequency spectrum contains lines at the two sideband frequencies $(n \pm q)f_{rev}$ of each revolution harmonic nf_{rev} (By courtesy of CERN, (©) CERN)

Using cosine functions to express I_s and x, the transverse 'dipole moment' of the beam current may also be written as

$$
I_s \cdot x = \left(ef_{rev} + 2ef_{rev_n} \sum_{m=1}^{\infty} \cos\big(m\omega_{rev}(t - t_p)\big) \right) \cdot \tilde{x} \cos(\omega_{rev} Qt + \mu_0)
$$

$$
= ef_{rev}\tilde{x} \left(\sum_{m=0}^{\infty} \cos\big(\omega_{rev}(m + Q)t - m\omega_{rev}t_p + \mu_0\big) \right.
$$

$$
\left. + \sum_{m=1}^{\infty} \cos\big(\omega_{rev}(m - Q)t - m\omega_{rev}t_p - \mu_0\big) \right) \tag{4.15}
$$

The pickup output voltage includes $Z'_{PU}(\omega)$ with $\omega = (m \pm Q)$ under the sums. This specifies the transverse signals of a single particle (see Fig. 4.4). To work out the Schottky noise we assume, once more, a coasting beam with random longitudinal particle positions $0 \le \omega_{rev}t_p \le 2\pi$, random 'betatron phases' $0 \le \mu_0 \le 2\pi$, single out a group of $N' = dN/d\omega$ particles and assume a distribution $n'(\tilde{x})$ of betatron amplitudes. We may interpret $N'(\omega)$ as the distribution of particles versus the sideband frequencies $\omega_m = (m \pm Q)\omega_{rev}$. Take the distributions uncorrelated (thus neglecting the dependence of Q and/or ω_{rev} on betatron amplitude) and proceed, as in the longitudinal case, to evaluate the spectral density of the mean squared dipole moment $D^2 = (I \cdot x)^2$ in the bands. One finds:

$$\frac{dD^2_{m\pm q}}{d\omega} = \frac{dN}{d\omega} \frac{e^2 f^2_{rev}}{2} \tilde{x}^2_{rms} \tag{4.16}$$

Here the r.m.s. amplitude comes from averaging $\tilde{x}^2$ and a factor $1/2$ from averaging the $\cos^2$ terms, Eq. (4.15). The width of the $n \pm q$ band is given by the frequency spread $\Delta\omega_{rev} = -\eta\omega_{rev} \cdot \Delta p/p$ and the chromatic q-spread $\Delta q = \xi Q \cdot \Delta p/p$

$$\Delta\omega_n = \Delta\big[(n \pm q)\omega_{rev}\big] = (n \pm q)\Delta\omega_{rev} \pm \Delta q \omega_{rev}$$

$$= \big[-\eta(n \pm q) \pm \xi Q\big]\omega_{rev}\Delta p/p \tag{4.17}$$

$$\Delta\omega_n \approx n\Delta\omega_{rev} = |\eta| n\omega_{rev}\Delta p/p \quad \text{for large } n$$

The chromaticity $\xi = (dQ/Q)/(dp/p)$ and the off-momentum parameter $\eta = (-d\omega_{rev}/\omega_{rev})/(dp/p) = \gamma_{tr}^{-2} - \gamma^{-2}$ (also called phase slip factor) are the usual 'machine constants'.

Integrating Eq. (4.16) over one separated band, the noise per band is found:

$$D^2_{n\pm q} = N\frac{e^2 f^2_{rev}}{2}\tilde{x}^2_{rms} \tag{4.18}$$

Approximating the shape of a band by a rectangle (of total width $\Delta\omega_n$) the noise density is

$$\frac{dD^2_{n\pm q}}{d\omega} = N\frac{e^2 f^2_{rev}}{2\omega_{rev}}\tilde{x}^2_{rms}\frac{\omega_{rev}}{\Delta\omega_n} = N\frac{e^2 f_{rev}}{4\pi}\tilde{x}^2_{rms}\frac{\omega_{rev}}{\Delta\omega_n} \tag{4.19}$$

Equation (4.19) gives (in the approximation of rectangular bands) the noise as long as bands are separated. When adjacent $n \pm q$ bands overlap the density is up to 2 higher. At high frequencies, where many bands overlap, the density is

$$\frac{dD^2}{d\omega} = N\frac{e^2 f_{rev}}{2\pi}\tilde{x}^2_{rms} \tag{4.20}$$

The factor of 2 comes in because the $n + q$ and $n - q$ bands contribute.

A spectrum analyzer, branched on a transverse pickup usually records the square root of the signal (4.16) multiplied by the pickup impedance $Z'_{PU\perp}$. Such transverse 'Schottky scans' are widely used for diagnostics. A practical example is Fig. 4.5, where the spectrum of the signal from a horizontal pickup is shown. The frequency band is centered around a revolution harmonic nf_{rev} and covers a width of f_{rev} so that the two sidebands, $(n \pm q)f_{rev}$, are visible. The height of these sidebands (with the electronic noise subtracted) is proportional to the r.m.s. betatron amplitude $\tilde{x}_{rms}$ and can thus be conveniently used to observe the transverse cooling process. Also: the non-integer part (q) of the tune can be determined from the distance of the sidebands. Finally—from a scan at lower harmonics where $n\Delta\omega_{rev} \le \omega_{rev}\Delta q$—the chromaticity (chromatic q-spread) can be deduced analyzing the width of adjacent $n \pm q$-bands, using Eq. (4.17).

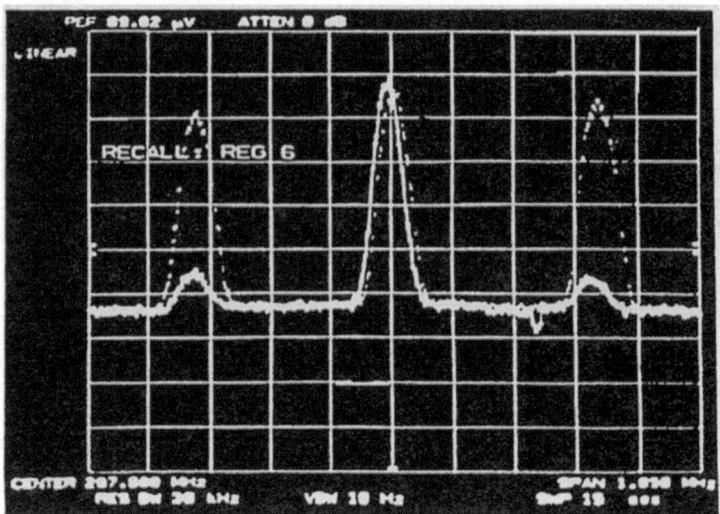

Fig. 4.5 A horizontal Schottky scan in LEAR at 600 MeV/c. The central band, the harmonic $n = 100$ of the revolution frequency, is visible as the beam is not completely centred at the position pickup. The right and left bands are the sidebands $(98 + Q)$ and $(102 - Q)$ where $Q \approx 2.3$. The difference between the base line of the trace and the bottom line (zero signal) is given by the noise of the pickup system. The span covers (approximately) an interval of frequency, f_{rev}. During cooling the height of the sidebands decreases (By courtesy of CERN, (©) CERN)

4.2 Cooling by Harmonics

Suppose now that the cooling loop acts proportionally to the error x of a single particle times the longitudinal current, i.e. it is convenient to work in terms of

$$y = x \cdot I(t) \tag{4.21}$$

where $I(t)$ consists of the single-particle current [Eq. (4.1)] plus the Schottky noise current. To fix our ideas, we may think of Palmer cooling, where the pickup signal is given by the particle current and the displacement $x = D \cdot \delta p/p$ at the pickup (assuming $x_\beta \ll D \cdot \delta p/p$).

Now transform to the frequency-domain, introduce the longitudinal Schottky noise (y_{Sch}) due to the other particles and assume for the moment a bandwidth $W = f_{rev}$ including only one band (n). The single passage correction is

$$y_c - y = -\Lambda_n y - a \cdot \Lambda_n y_{Sch} \tag{4.22}$$

where $y = y(\omega)$; Λ_n is the 'gain' in the band under consideration and 'a' is a constant which describes the response of the particle to the Schottky noise. Rather than derive this response, we will adjust 'a' to recover results of the time-domain analysis. We invoke that a particle will mainly respond to components of the noise with a frequency very close to its revolution harmonics. Other frequency components will

rapidly fall out of synchronism with the particle. The perturbation is then proportional to the noise density at the corresponding frequency nf_{rev}.

As before we work out the change per turn of the squared error of a test particle

$$\Delta(y^2) = y_c^2 - y^2 = -2\Lambda_n y^2 + \Lambda_n^2 y^2 + 2(1 - \Lambda_n)ya\Lambda_n y_{Sch} + (a\Lambda_n y_{Sch})^2$$

Average over many turns, take the noise and test-particle signal uncorrelated so that
$\langle y \cdot y_{Sch} \rangle = 0$, and anticipate $\Lambda_n \ll 1$ (see below), i.e. neglect $\Lambda_n^2 y^2 \ll 2\Lambda_n y^2$ to
obtain the cooling rate:

$$\frac{1}{\tau_{x^2}} = f_{rev}\frac{-\langle\Delta(y^2)\rangle}{\langle y^2\rangle} = f_{rev}\left(2\Lambda_n - \frac{a^2\Lambda_n^2\langle y_{Sch}^2\rangle}{\langle y^2\rangle}\right) \tag{4.23}$$

As a 'trial' for $\langle y_{Sch}^2 \rangle$ we take the power per band [Eq. (4.9)] multiplied with x_{rms}^2
and with the factor $\omega_{rev}/(n\Delta\omega_{rev})$ appearing in Eq. (4.10) for separated bands. As
we reserve 'a' to be adjusted in accordance with the time-domain results we have
the freedom of this choice. Hence

$$\langle y_{Sch}^2 \rangle = x_{rms}^2 2Ne^2 f_{rev}^2 \begin{cases} \omega_{rev}/(n\Delta\omega_{rev}) & \text{separated bands } (n < \omega_{rev}/\Delta\omega_{rev}) \\ 1 & \text{overlapping bands } (n > \omega_{rev}/\Delta\omega_{rev}) \end{cases} \tag{4.24}$$

Taking again as a typical particle one with $y = y_{rms}$ we write [see Eqs. (4.1)
and (4.21)]

$$y^2 = x_{rms}^2 (2ef_{rev})^2 \tag{4.25}$$

Rewrite Eq. (4.23):

$$\frac{1}{\tau_{x^2}} = f_{rev}\left[2\Lambda_n - \Lambda_n^2\frac{a^2}{2}NM_n\right]$$

$$\text{with} \quad M_n = \begin{cases} \omega_{rev}/(n\Delta\omega_{rev}) & \text{separated bands} \\ 1 & \text{overlapping bands} \end{cases} \tag{4.26}$$

Identify $\Lambda_n = 2g_n/N$ and $a = 1$, to recover the familiar result [e.g. Eq. (2.21) of
Chap. 2 for $W = f_{rev}$] in the limit of large $n\Delta\omega_{rev}$, i.e. for good mixing:

$$\frac{1}{\tau_{x^2}} = \frac{2f_{rev}}{N}(2g_n - g_n^2) \quad \text{(one band and } M_n = 1) \tag{4.27}$$

For $\ell = W/f_{rev}$ bands we add up the contributions (4.27) assuming that the
'cross terms' between different bands average out. Including finite frequency spread
($M_n > 1$), and turning to the cooling time for x instead of x^2 we have:

$$\frac{1}{\tau} = \frac{f_{rev}}{N}\sum_{n_1}^{n_1+\ell}(2g_n - g_n^2 M_n) = \frac{W}{N}\cdot\frac{1}{\ell}\sum_{n_1}^{n_1+\ell}(2g_n - g_n^2 M_n) \tag{4.28}$$

Equation (4.28), which expresses the cooling rate by an average $[\frac{1}{\ell}\sum(2g_n - g_n^2 M_n)]$ over the pass band, is the main result of this subsection. For constant $g_n = g$ (independent of n) it simplifies to

$$\frac{1}{\tau} = \frac{W}{N}\left(2g - g^2\frac{1}{\ell}\sum_{n_1}^{n_1+\ell} M_n\right) \tag{4.29}$$

We then identify the mixing factor (in the approximation of constant g_n):

$$M = \frac{1}{\ell}\sum_{n_1}^{n_1+l} M_n \tag{4.30}$$

If all bands are non-overlapping $[M_n = \omega_{rev}/(n\Delta\omega_{rev})$ for all n involved], Eq. (4.30) becomes

$$M = \frac{f_{rev}}{W}\frac{\omega_{rev}}{\Delta\omega_{rev}}\sum_{n_1}^{n_1+\ell}\frac{1}{n} \approx \frac{f_{rev}}{W}\frac{\omega_{rev}}{\Delta\omega_{rev}}\ln(f_{max}/f_{min}) \tag{4.31}$$

Using the sample length $T_s = 1/(2W)$ (see Eq. (2.1), Chap. 2), and $|\Delta\omega_{rev}/\omega_{rev}| = |\Delta T_{rev}/T_{rev}|$ one finds:

$$\frac{f_{rev}}{W}\frac{\omega_{rev}}{\Delta\omega_{rev}} = \frac{T_s}{\Delta T_{rev}/2}$$

Equation (4.31) then agrees with the common sense expectation that a spread in revolution times equal to about one sample length assures good mixing ($M \to 1$).

The frequency-domain analysis has thus provided us with more insight as we obtain:

- A more general cooling equation which includes gain variation with n and mixing.
- It can be optimised, harmonic by harmonic.
- An interpretation of bad mixing as enhancement of the incoherent heating because the Schottky noise density is increased by $M_n > 1$.

This is the interpretation of mixing introduced by Sacherer in 1978 [40] shortly after an expression similar to Eq. (4.30) had been derived by van der Meer [41] using time-domain analysis. The approximation of 'rectangular' frequency distribution implicitly used in this subsection is slightly pessimistic if we are mainly interested in the cooling speed of particles with large error. This is because in the tails the Schottky noise density is lower than average.

A remark is now in order: In writing Eq. (4.22) we have tacitly assumed that there are now phase-shifts between the particle and its self-correction, in addition to the time-of-flight errors to be analysed in Sect. 4.4. In reality the Λ_n are complex functions (amongst other things due to the pickup and kicker response) and the coherent and incoherent effect are determined by $\text{Re}(\Lambda_n)$ and $|\Lambda_n|^2$ respectively as will become clearer in Chaps. 5 and 7. This reduces the coherent effect. The results of the present section are approximations as they do not include phase-shifts in the cooling loop. Also an additional effect, known as feedback via the beam (to be treated in Chap. 7), is so far not included. In Appendix B we express the gain g_n by pickup and kicker impedance and amplification.

4.3 Noise-to-Signal Ratio by Harmonics

The electronic noise entering the low level stage of the system has an important influence on stochastic cooling. There are different sources of this noise, including the pickup system and its terminations, as well as cables, combiners and the amplifier itself. Frequently, when the beam intensity is low one has to cryogenically cool part or all of these elements to obtain a usable signal to noise ratio and/or acceptable power requirements for the amplifier.

It has become customary to refer the noise to the input of the amplifier system and characterise it by a noise index ν (db). The noise voltage is assumed to have a constant ('white') spectral power density. For an input impedance R_i its value is

$$\frac{dV_\nu^2}{d\omega} = \frac{10^{\nu/10}}{2\pi} k T_a R_i \approx \left(7.0 \times 10^{-22} \text{ W s}\right) \cdot 10^{\nu/10} R_i \qquad (4.32)$$

In Eq. (4.32) $k = 1.38 \times 10^{-23}$ W s/K is the Boltzmann constant and $T_a = 320$ K the ambient temperature. A typical noise index for non-cryogenic systems is 3 db. Alternatively the noise density is characterised by an equivalent noise temperature T_{eq}:

$$\frac{dV_\nu^2}{d\omega} = \frac{1}{\pi} k T_{eq} R_i \qquad (4.33)$$

which is obtained by inserting $2T_{eq}$ instead of $T_a 10^{\nu/10}$ in Eq. (4.32).

As an example:

$$\nu = 1.5 \text{ db} \quad (T_{eq} = 226 \text{ K}) \quad R_i = 50 \ \Omega \rightarrow \frac{dV_\nu^2}{d\omega} = 4.96 \times 10^{-20} \text{ V}^2\text{/Hz}$$

$$\Delta V_\nu^2 = (17.6 \ \mu\text{V})^2 \quad \text{for a band width of } \Delta\omega = 2\pi \cdot 1 \text{ GHz}$$

The total Schottky noise and electronic noise (the latter assumed to be 'white') entering the cooling loop is sketched in Fig. 4.6.

To work out the noise-to-signal ratio U, we compare the electronic noise, Eq. (4.32), with the spectral density of the Schottky noise. In the definition of U, good mixing has been assumed, so we have to take Eq. (4.11), for the comparison. For a longitudinal pickup the output voltage is $V_{pu} = Z_{PU}//I_{beam}$ (see Chap. 3). A somewhat more detailed record of noise transmission will be given in Sect. 5.7 below. We anticipate here, that noise transmitted through a linear network with transfer

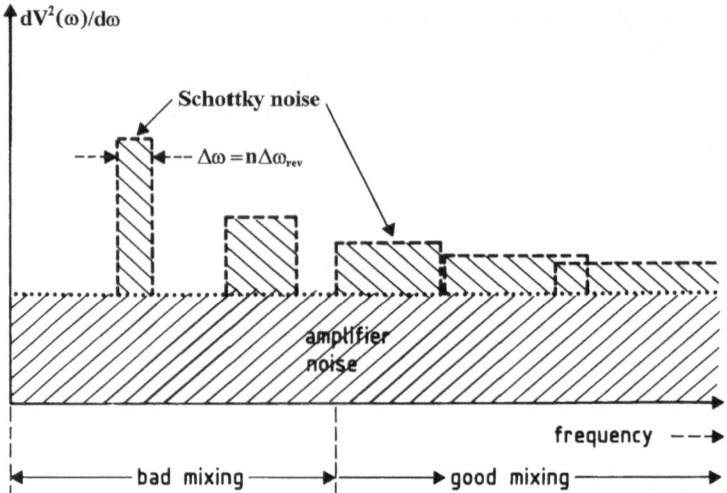

Fig. 4.6 Amplifier and Schottky noise which determine the incoherent heating. The amplifier noise has a continuous spectrum with a density given by the noise index v of the low level system. The Schottky noise has the band spectrum (By courtesy of CERN, (©) CERN)

impedance $Z(\omega)$ transforms according to

$$\frac{dV^2}{d\omega} = |Z(\omega)|^2 \frac{dI^2}{d\omega} \tag{4.34}$$

Hence referring the electronic and Schottky noise to the to the pickup output we have

$$U = \frac{dV_v^2}{|Z_{PU//}^2|dI_{Schottky}^2} = \frac{10^{v/10}kT_aR_i}{|Z_{PU//}^2|2e^2f_{rev}N} \tag{4.35}$$

Here $Z_{PU//}$ is the total pick up transfer impedance, Chap. 3. For loop couplers with combiners one has $Z_{PU//}^2 = n_p Z_{P//}^2$ when n_p units with an impedance $Z_{P//}$ are added up. Note also, that $Z_{PU//}$ and hence $U = U(\omega)$ is usually frequency dependent and varies over the cooling band. As an example:

$$N = 10^9, \quad f_{rev} = 1\,\text{MHz}, \quad n_p = 25, \quad Z_{p//} = 25\,\Omega$$

$$\left(\text{loop at midband, } F(\omega) = 1, \, g_{//} = 1\right)$$

$$\rightarrow Z_{PU//} = 125\,\Omega, \quad v = 3\,\text{db}, \quad T_a = 320\,\text{K}, \quad R_i = 125\,\Omega$$

$$\rightarrow U \approx 1.4 \quad \text{at midband, increasing with } F(\omega)^{-2},$$

$$\text{Eq. (3.17), towards the band edges}$$

For a transverse pickup system the output voltage $V_{PU} = Z'_{PU\perp} x I_{beam}$ has to be taken as the signal voltage. Defining the noise-to-signal ratio with respect to the

Schottky noise for complete overlap, Eq. (4.20), one obtains

$$U_\perp = \frac{10^{\nu/10} k T_a R_i}{|Z'^2_{PU\perp}| \tilde{x}^2_{rms} e^2 f_{rev} N}$$ (4.36)

Here $U_\perp$ increases as the beam shrinks unless movable pickup plates are used with a distance that shrinks with the beam width $\tilde{x}_{rms}$. Note that both U and $U_\perp$ are proportional to $1/N$. Therefore the optimum cooling rate in the noise limit (Eq. (2.31) of Chap. 2 for $U \gg M$) is independent of the intensity. This explains the 'amplifier noise limit' ('levelling-off' of the cooling time vs. N curves at low N) in Fig. 2.12.

Above we have assumed that the gain can (slowly) vary with the harmonic number but is (practically) constant over the width of each band. This is a valid approximation for betatron and Palmer cooling unless steep gain shaping filters inserted into the cooling loop. For momentum cooling by the filter method both Schottky and electronic noise are suppressed at the nominal harmonics of the revolution frequency and hence the influence on small $\Delta p/p$ particles is much reduced.

4.4 Mixing Pickup to Kicker

Let the 'typical' test particle pass the pickup at $t = 0$ and the kicker at $t = t_{PK} + \Delta t_{PK}$. Let the correction take the nominal time t_{PK} to travel from pickup to kicker. Assume again a bandwidth $W = f_{rev}$ covering one harmonic (n) for the moment, and regard only the effect of the particle upon itself. From Eqs. (4.21) and (4.22) the single passage correction is written as

$$y_c = y - \lambda_n \cdot x \cdot I_s(\Delta t_{PK})$$ (4.37)

For simple systems such as betatron or Palmer cooling we can assume, $I_s(\Delta t_{PK}) \propto \cos(n\omega_{rev}\Delta t_{PK})$. Going through the standard procedure to work out $\langle y_c^2 - y^2 \rangle$, neglecting once again other phase-shifts in the cooling loop, summing over harmonics we find the 'coherent effect', expressed by the first term in Eq. (4.28), modified as:

$$\frac{1}{\ell} \sum_{n1}^{n1+\ell} 2g_n \rightarrow \frac{1}{\ell} \sum_{n1}^{n1+\ell} 2g_n \cdot \cos(n\omega_{rev}\Delta t_{PK})$$ (4.38)

Hence, the harmonic method gives an improved expression also for the unwanted mixing, and we can identify the factor $(1 - \tilde{M}^{-2})$ used in Chap. 2,

$$g \cdot (1 - \tilde{M}^{-2}) = \frac{1}{\ell} \sum_{n1}^{n1+\ell} g_n \cdot \cos(n\omega_{rev}\Delta t_{PK})$$ (4.39)

as the average cosine of the phase error in the pass-band. The incoherent effect remains unchanged, in our approximation as the Schottky and electronic noise are stationary and thus independent of the time of arrival.

Fig. 4.7 Reduction of
coherent correction (mixing
pickup to kicker) as a
function of $\delta p/p$ of a
particle. Band-width of 1
octave, constant gain.
Cooling only works for the
particles with $1 - \tilde{M}^2 > 0$

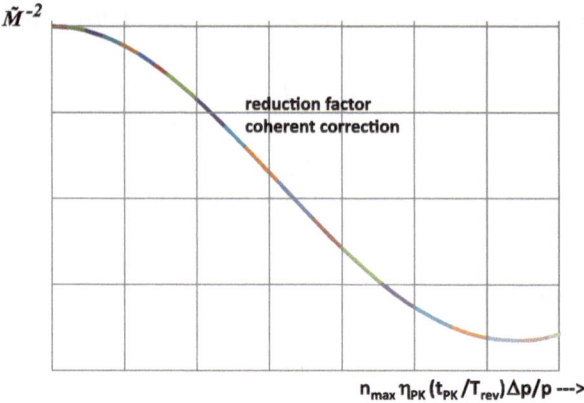

Equation (4.38) sets limit for $n_{\max}|\eta_{PK}|\delta p/p$. In fact the argument of the co-
sine in Eq. (4.38) may be written as $n\omega_{rev}\Delta t_{PK} = 2\pi n|\eta_{PK}|(t_{PK}/T_{rev})\delta p/p$.
Here t_{PK}/T_{rev} is the flight time pickup to kicker in relation to the revolution time;
$|\eta_{PK}| = (\Delta t_{PK}/t_{PK})/(\delta p/p)$ is the local slip factor [where in a regular storage
ring: local $\approx$ whole-ring slip factor, i.e. $|\eta_{PK}| \approx |\eta| = (\Delta T_{rev}/T_{rev})/(\delta p/p)$]. De-
tails of the $\delta p/p$-limit depend on the gain profile and the band-width. Frequently
one requires

$$n_{\max}|\eta_{PK}|(t_{PK}/T_{rev})\delta p/p < 0.25 \qquad (4.40)$$

In fact, Fig. 4.7 shows the coherent effect, Eq. (4.38), as a function of
$n_{\max}|\eta_{PK}|(t_{PK}/T_{rev})\delta p/p$. In this example a band of one octave ($n_{\max} = 2n_{\min}$)
and constant gain is assumed. One concludes that a particle with a $\delta p/p$ such that
$n_{\max}|\eta_{PK}|(t_{PK}/T_{rev})\delta p/p = 0.25$ receives already less than half the nominal correction. For values larger than 0.34, cooling of the particle becomes impossible (as
the coherent effect is negative). For the example: $t_{PK}/T_{rev} = 0.5$, and $\eta_{PK} \approx \eta$ (di-
agonal signal path, local and whole-ring slipping factor equal) Eq. (4.40) simplifies
to

$$n_{\max}|\eta|\delta p/p < 0.5 \qquad (4.41)$$

This exemplifies the momentum spread limit for Palmer-Hereward and transverse
cooling. We now return briefly to the filter method which was already treated in
Sect. 2.14: In this case the limit on $n_{\max}|\eta|\delta p/p$ is more stringent (typically by a
factor of 3). The signals are compared near the pickup with those from the previous
turn to which the traveling time to the kicker adds. It is then essentially $\eta_{eff} \equiv$
$\eta + \eta_{PK}(t_{PK}/T_{rev})$ instead of $\eta_{PK}(t_{PK}/T_{rev})$ that enters into the limit (4.40). For
more details see Appendix A.

Once more, like in Sect. 4.2, we have neglected additional phase-shifts (e.g. due
to pickup and kicker response) in the coherent term. Thus results of this subsection
are still approximate but exemplify the tendency. Having improved the time-domain
results of momentum cooling, we now turn to a more detailed analysis of betatron
oscillation cooling.

Chapter 5
A More Detailed Derivation of Betatron Cooling

5.1 Betatron Equation

Before entering into details, it is worth trying to establish a simple picture of beta-tron cooling in which the various phenomena can be identified. Consider first the smooth sinusoidal approximation for the betatron motion [3, 4] of a single particle in a storage ring, with forcing terms on the right-hand side arising from its proper motion, the motion of other particles (subscript j) and system noise:

$$\ddot{x}(t) + \omega_\beta^2 \cdot x(t) = \underbrace{G \cdot x(t - t_{PK})}_{\substack{\text{coherent effect} \\ \text{- mixing} \\ \text{PICKUP to } K \\ \text{- betatron phase error}}} + \underbrace{\sum_n G_{ij} \cdot x_j(t - t_{PK})}_{\substack{\text{incoherent effect} \\ \text{- mixing} \\ K \text{ to PICKUP} \\ \text{- signal shielding}}} + \underbrace{\text{'system noise'}}_{\substack{\text{additional incoh. effect} \\ \text{- enhancement of cooling}}}$$

(5.1)

We interpret the left-hand side as the motion on entering the cooling kicker (K) and the forcing terms on the right-hand side as being derived from the motion seen earlier (i.e. at $t - t_{PK}$) in the pickup. The characteristics of the pickup, amplifier, transmission system and kicker enter into the coefficients G, G_{ij} and into the 'system noise'; $\omega_\beta = Q\omega_{rev}$ is the betatron frequency given by the tune (Q) of the storage ring and the particle revolution frequency.

If we neglect the incoherent terms in Eq. (5.1), we obtain a single-particle cooling equation,

$$\ddot{x}(t) + \omega_\beta^2 \cdot x(t) = G \cdot x(t - t_{PK})$$

(5.2)

We expect a solution of the form: $x(t) = \tilde{x}e^{i\omega t}$, $x(t - t_{PK}) = \tilde{x}e^{i(\omega t - \Delta\psi)}$ with $\Delta\psi = \omega_\beta(t - t_{PK})$. Substituting into Eq. (5.2) gives:

$$-2\omega_\beta \Delta\omega \approx \omega_\beta^2 - \omega^2 = G \cdot e^{-i\Delta\psi}$$

(5.3)

This is the expected response of a feedback system. The real part of $\Delta\omega$ is the frequency shift of the perturbed oscillation and the imaginary part gives the damping

D. Möhl, *Stochastic Cooling of Particle Beams*, Lecture Notes in Physics 866, DOI 10.1007/978-3-642-34979-9_5, © Springer-Verlag Berlin Heidelberg 2013

(or heating) of the oscillation.

$$\frac{1}{\tau} = \text{Im}(\Delta\omega) = \text{Re}\left\{\frac{G \cdot e^{i(\Delta\psi - \pi/2)}}{2\omega_\beta}\right\} \tag{5.4}$$

Equation (5.4) would be exact (in smooth approximation) if the observation and the feedback on the beam would be all around the ring, which is manifestly not the case. We will now, therefore, investigate the effects of localised observation and correction and also try to remove the smooth approximation.

5.2 Orbit Equation for a Localised Kick

The orbit deviation $(x = x_\beta)$ in a storage ring with constant kicks is given by the well-known equation [3, 4],

$$x''(s) + K(s) \cdot x(s) = \left(\frac{\Delta B_z(s)}{B_0\rho_0} + \frac{eE_x(s)}{\beta pc}\right) \equiv F(s) \tag{5.5}$$

where:

$K(s)$ is the focusing function [m^{-2}],
E_x the transverse electric field [V/m],
ΔB_z the transverse magnetic field [T] additional to the guide field (vertical field for horizontal orbit deformation),
$B_0\rho_0$ the magnetic rigidity [T m] of the particle, 3.3356×10^{-9} [T m] $p/$[eV/c] for protons,
$p = \beta\gamma m_0 c$ the particle momentum [eV/c],
$\beta = v/c, \gamma$ the relativistic factors,
$s = \beta ct$ the distance along the orbit [m],
$x''(s)$ the second derivative of the orbit deviation with respect to s.

Using the well known Courant and Snyder transformation [3, 4] the equation can be rewritten as a 'driven harmonic oscillator' with fixed frequency Q rather than with azimuthally varying $K(s)$:

$$\eta_\beta''(\phi) + Q^2\eta_\beta(\phi) = Q^2\beta_x^{3/2}F(s) \tag{5.6}$$

Here $\eta_\beta(s) = x\beta^{-1/2}$ is the normalised displacement, $d\phi = ds/(Q\beta)$ defines the Courant and Snyder angle which increase by 2π per turn, $\beta_x(s)$ is the betatron amplitude function of the storage ring and the dash ($'$) now indicates differentiation with respect to ϕ.

Equation (5.6) is quite general, but we are especially interested by single narrow kicks of length Δs, which we can represent by a delta function, periodic with the revolution frequency:

$$F(s) = F_{(s=s_0)}\Delta s \sum_{n-\infty}^{\infty} \delta(s_0 - 2\pi nR) \tag{5.7}$$

Equation (5.7) is a good representation of a short kicker providing a constant kick, but analytically it would be easier to manipulate, if we could replace the discontinuous delta functions with a continuous function. This can readily be done by making a Fourier expansion of the quantity $\beta_x^{3/2} F$ which appears on the r.h.s. of Eq. (5.6):

$$\beta_x^{3/2}(s) \cdot F(s) = \sum_{\ell=-\infty}^{\infty} f_\ell e^{i\ell\phi} \tag{5.8}$$

where,

$$f_\ell = \frac{1}{2\pi} \int_0^{2\pi} \beta_x^{3/2} F e^{-\ell\phi} d\phi = \frac{1}{2\pi} \int_0^{2\pi R} \beta_x^{1/2} F e^{-\ell\phi(s)} ds / Q$$

Since F is a delta function, the integral simply leads to Fourier coefficients which are all equal (accepting that in the Fourier expansion of Eq. (5.6) very high frequency components, with $f > c/\Delta s$ are not required):

$$f_\ell = \frac{F\Delta s}{2\pi Q} \beta_K^{1/2} \equiv \frac{\theta}{2\pi Q} \beta_K^{1/2} \tag{5.9}$$

Here β_K is the value of the beta function at the kicker and

$$\theta = \frac{\Delta p_\perp}{p} = \left(\frac{\Delta B_z}{B_0 \rho_0} + \frac{eE_x}{\beta p c} \right) \Delta s = F\Delta s \tag{5.10}$$

is the kick angle already introduced in Eq. (3.6) of Sect. 3.1.

We can now rewrite Eq. (5.6) in terms of a continuous function

$$\eta_\beta''(\phi) + Q^2 \eta_\beta(\phi) = \frac{Q}{2\pi} \beta_K^{1/2} \theta \sum_{\ell=-\infty}^{\infty} e^{i\ell\phi} \tag{5.11}$$

This is in fact all we really need, but we can make two variable changes, which will make the equation more familiar. Firstly, since we prefer to think in terms of time, we can introduce a 'time-like' variable ϕ/ω_{rev} i.e. the Courant and Snyder phase, ϕ, scaled by the revolution angular frequency, ω_{rev}. Rather loosely we will refer to this variable as t. In fact, this lack of rigour is not too serious, since t will coincide with the true time at least once every revolution at the kicker ($s = 0$), which is the one point where true time is important. In addition, in most lattices ϕ/ω_{rev} will not stray far from true time at any point in the ring. Secondly, we like to think in transverse deviation x, so we undo the normalisation of the variable η_β. We are only really interested in true deviations at the pickup, so we define a position variable $\eta_\beta \beta_{PU}^{1/2}$ which gives true position once per turn at the pickup and again we loosely call it x:

$$t = \frac{\phi}{\omega_{rev}} \approx \text{true time}, \qquad x = \eta_\beta \beta_{PU}^{1/2} \approx \text{true position} \tag{5.12}$$

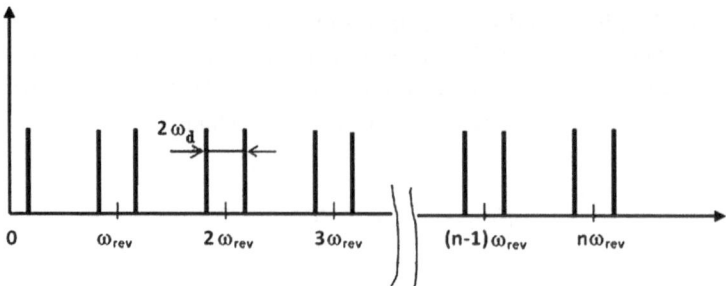

Fig. 5.1 Frequency spectrum seen by a particle when a short kicker is excited with a sinusoidal voltage of frequency ω_d (By courtesy of CERN, (©) CERN)

Using (5.12) in (5.11), we find

$$\ddot{x}(t) + Q^2 \omega_{rev}^2 \cdot x(t) = \frac{Q \omega_{rev}^2 \beta_K^{1/2} \beta_{PU}^{1/2}}{2\pi} \theta \sum_{\ell-\infty}^{\infty} e^{i\ell\omega_{rev}t} \qquad (5.13)$$

Thus the betatron motion is driven by an infinite set of Fourier harmonics separated, one from the other, by the revolution frequency ω_{rev}.

If we now modulate the kick we can still use the expansion (5.8). We simply have to make the kick strength θ of Eq. (5.9) a function of time. The result is that all Fourier coefficients f_ℓ (Eq. (5.9)) are modulated by the same factor. This leads to the phenomenon known as transverse RF knockout. Let us rewrite Eq. (5.13) as:

$$\ddot{x}(t) + \omega_\beta^2 \cdot x(t) = \frac{Q \omega_{rev}^2 \beta_K^{1/2} \beta_{PU}^{1/2}}{2\pi} \tilde{\theta} \sum_{\ell=-\infty}^{\infty} e^{i(\ell\omega_{rev}+\omega_d)t} \qquad (5.14)$$

$\tilde{\theta}$ being the amplitude of the kick, $\theta = \tilde{\theta}e^{i\omega_d t}$.

Equation (5.14) is expressed with negative and positive frequencies, which correspond to the slow and fast waves set up by the disturbing kick. In fact above we assumed a complex excitation of the kicker as this simplifies the algebra. In real life we deal with cosine rather than $e^{i\omega_d t}$-type of kicker fields. To go from the complex to the real world one has to take the real parts of Eq. (5.14). Then the r.h.s. contains terms which can be written in the form $\cos(|\ell|\omega_{rev} + \omega_d)t$, and $\cos(-|\ell|\omega_{rev} + \omega_d)t = \cos(|\ell|\omega_{rev} - \omega_d)t$. Thus the particle 'sees' the frequencies $|\ell|\omega_{rev} \pm \omega_d$ i.e. two sidebands spaced by the kicker excitation frequency left and right of each revolution harmonic. This is illustrated in Fig. 5.1, where the spectrum of the excitation [r.h.s. of Eq. (5.14)] is sketched.

The revolution sidebands at $|\ell|\omega_{rev} \pm \omega_d$ are very similar to the sidebands at $\omega_{carrier} \pm \omega_{mod}$ of an amplitude-modulated oscillator.

If the driving term in Eq. (5.14) has a harmonic at the response frequency, ω_β, then the beam will behave resonantly. The resonance condition is $(\ell\omega_{rev} + \omega_d)^2 =$

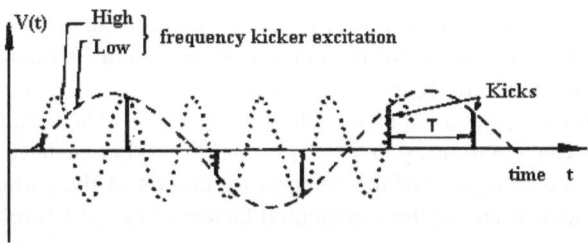

Fig. 5.2 Example of how a beam can be driven in the same way by different frequencies applied from a short kicker. The *bars* are the kicker voltages at the moment the particle passes, i.e. the kicks experienced. The low frequency (ω) and the higher frequency ($\omega + n\omega_{rev}$) excitation produce the same apparent kicks (By courtesy of CERN, (©) CERN)

ω_β^2. This yields resonances for

$$\omega_d = |\ell|\omega_{rev} \pm \omega_\beta = \left(|\ell| \pm Q\right)\omega_{rev} = \left(\underbrace{|\ell| \pm \|Q\|}_{n} \pm q\right)\omega_{rev} \tag{5.15}$$

Here (as in Sect. 4.1) $\|Q\|$ is the integer part of the tune, q is the fractional part and, $n = |\ell| \pm \|Q\|$ is also an integer. Equation (5.15) shows that the beam will respond resonantly at the 'betatron sidebands' $(n \pm q)\omega_{rev}$ centred on the revolution frequency harmonics. Take as an example LEAR at 600 MeV/c with $f_{rev} = 1$ MHz, $q = 0.3$. Resonant beam response ('RF-knockout') will occur, when the kicker is excited at $(0 + q)f_{rev} = 0.3$ MHz, or $(1 - q)f_{rev} = 0.7$ MHz, or $(1 + q)f_{rev} = 1.3$ MHz etc.

Figure 5.2 sketches how two voltages of different frequency (ω and $\omega + n\omega_{rev}$ respectively) on a short kicker can produce the same series of kicks as seen by a particle. Note the analogy to an RF accelerating cavity which can in principle work at any revolution harmonic. Note also the similarity with the reciprocal of the present problem namely the response of the pickup to the betatron oscillation of a particle as treated in Sect. 4.1 before. The interaction frequencies, Eq. (4.14), are exactly the same as for the kicker response, Eq. (5.15): the beam 'talks' and 'responds' at the same frequencies.

5.3 Coherent Effect

We shall now use Eq. (5.14) to have a fresh look at the motion of particle 'i'. We retain for the moment only its 'self-terms'. We take the kick $\theta = eV_\perp/(\beta pc)$ [Eq. (3.6)] on the r.h.s. of Eq. (5.14) given by the pickup signal Eq. (4.12) of the circulating particle transmitted through the cooling loop onto the kicker. Proper account has to be taken of the time of flight as well as of delays and phase-shifts in the cooling loop. A point to retain is that all signal transmission occurs at the $(m + Q)\omega_{rev}$ betatron sidebands (with $m = -\infty$ to ∞) or equivalently at the $(n \pm q)\omega_{rev}$ frequencies within the pass-band.

The particle takes a nominal time t_{PK} and the cooling signal a time t_c to travel from pickup to kicker. The electronic delay t_c is in general frequency dependent. We include the deviation from nominal by a phase factor $\varphi_m = m\omega_{rev}(t_c - t_{PK})$ in the representation of the cooling signal entering Eq. (4.12). The initial phase μ_0 can be absorbed into the betatron excursion, $x_i = \tilde{x}e^{i\omega_\beta t + \mu_0}$, but we have to include the phase-advance $\Delta\psi = \omega_\beta t_{PK}$ of the betatron oscillation of the particle on its way from pickup to kicker. Hence the exponential factors in Eq. (4.12) referred to at the kicker are written as

$$\exp[i\{\omega_m t - \Delta\psi + \varphi_m\}] \quad \text{with } \omega_m = (Q + m)\omega_{rev} \tag{5.16}$$

To complete the driving term in the betatron equation (5.14) we introduce a transfer function $G(\omega_m)$. It includes, via $\tilde{\theta}$, the pickup response $Z'_{P\perp}$ [Table 3.3 of Chap. 3], the transfer function of the cooling loop between pickup and kicker (with cables, amplifiers, filters, etc.) as well as the kicker response $K_\perp$ [Table 3.3 of the Chap. 3]. Let us for simplicity absorb the constant factor

$$\frac{Q\omega_{rev}^2 \beta_K^{1/2} \beta_{PU}^{1/2}}{2\pi} \tilde{\theta}/x_i$$

into this 'transfer function' but keep x_i separate. Hence we rewrite Eq. (5.14) as

$$\ddot{x}_i + \omega_\beta^2 x_i = x_i e^{-i\Delta\psi} \cdot \sum_{m=-\infty}^{\infty} |G(\omega_m)| \cdot e^{i(m\omega_{rev}t + \varphi_m)} \cdot \sum_{\ell=-\infty}^{\infty} e^{i\ell\omega_{rev}t} \tag{5.17}$$

We note that the transfer function "single particle dipole current at the pickup to kick" is $\sum_{m=-\infty}^{\infty} G(\omega_m)/ef_{rev}$.

The phase shifts of $G(\omega_m)$ are included in φ_m and the initial phase drops out as we take $x_i = \tilde{x}e^{i(\omega_\beta t + \mu_0)}$ on both sides of Eq. (5.17). The second sum in Eq. (5.17) is the 'sampling term'—appearing in Eq. (5.14)—due to the fact that the particle passes the short kicker once per turn. The first sum clearly is due to the localised nature of the pickup. Equation (5.17) is almost the same as Eq. (5.2) except that we now include the frequency dependent 'gain' and localised pickup and kicker. The product of the two sums in Eq. (5.26) can readily be converted into a double sum noting that

$$\sum_m a_m \cdot \sum_\ell b_\ell = \sum_m \sum_\ell a_m \cdot b_\ell \tag{5.18}$$

Hence the r.h.s. of Eq. (5.17) may be expressed as

$$x_i e^{-i\Delta\psi} \cdot \sum_{\ell=-\infty}^{\infty} \sum_{m=-\infty}^{\infty} G(\omega_m) \cdot e^{i(m+\ell)\omega_{rev}t + i\varphi_m} \tag{5.19}$$

Thus Eq. (5.17) may be interpreted as pertaining to an oscillator with a frequency shift that varies in time. An approximate solution to such equations is obtained by taking the time average of the frequency shift only, i.e. if we retain terms with $\ell = -m$ in Eq. (5.19) and drop the rapidly oscillating frequency shifts. Using this approximation Eq. (5.17) becomes:

$$\ddot{x}_i + \omega_\beta^2 x_i = x_i e^{-i\Delta\psi} \cdot \sum_{m=-\infty}^{\infty} \left| G(\omega_m) \right| \cdot e^{i\varphi_m} \tag{5.20}$$

This defines the change of betatron frequency:

$$\Delta\omega_\beta \approx \frac{-e^{-i\Delta\psi}}{2\omega_\beta} \cdot \sum_{m=-\infty}^{\infty} \left| G(\omega_m) \right| \cdot e^{i\varphi_m} \tag{5.21}$$

As we assumed $x = \tilde{x} e^{i(\omega_\beta t + \mu_0)}$, the damping rate is given by

$$1/\tau = \text{Im}(\Delta\omega_\beta) = \frac{1}{2\omega_\beta} \cdot \sum_{m=-\infty}^{\infty} \left| G(\omega_m) \right| \cdot \sin(\Delta\psi - \varphi_m) \tag{5.22}$$

Optimum cooling is obtained, if for all m the phase factor is properly chosen such that $\sin(\mu - \varphi_m) = 1$. This requires

$$\Delta\psi - \varphi_m = \pi/2 \quad (\text{modulo } \pi) \tag{5.23}$$

Usually Eq. (5.23) is satisfied minimising the phases φ_m and putting the kicker at the 'proper' betatron phase advance from the pickup, i.e. by choosing:

$$\varphi_m = 0 \quad \text{and} \quad \Delta\psi = \pi/2, 5\pi/2, 9\pi/2, \ldots$$
$$\text{or with signal inversion } \Delta\psi = 3\pi/2, 7\pi/2, \ldots \tag{5.24}$$

Ideally $\varphi_m = 0$ requires a signal delay of the cooling loop equal to the particle travelling time pickup to kicker, independent of frequency and this requires phase compensation over the whole pass-band. If the optimum betatron advance pickup to kicker is not possible one can in principle include filters (as proposed by Thorndahl) with a time delay characteristic $t_c(\omega)$ such that Eq. (5.23) is still satisfied. This requires however 'steep' filters with a delay varying very strongly from the $(n - q)$ to the neighbouring $(n + q)$ betatron band.

Further effects can be identified from Eq. (5.22):

• The time-of-flight error Δt_{PK} of a particle ('mixing between pickup and kicker') as well as improper delay Δt_c and additional phase-shifts of the cooling loop or improper pickup to kicker betatron phase advance. They all appear in the phase factors in Eq. (5.21). The equation may be re-written as

$$\frac{1}{\tau} = \frac{1}{2\omega_\beta} \operatorname{Re}\left(\sum_{m=-\infty}^{\infty} |G(\omega_m)| \cdot e^{i(\delta\varphi_m - \delta\Delta\psi)} \right)$$

$$= \frac{1}{2\omega_\beta} \sum_{m=-\infty}^{\infty} |G(\omega_m)| \cdot \cos(\delta\varphi_m - \delta\Delta\psi) \qquad (5.25)$$

with $(\delta\varphi_m - \delta\Delta\psi)$ the deviation from Eq. (5.23).

- The finite bandwidth of the system. It is expressed via the transfer function $G(\omega_m)$ which tends to zero for frequencies ($|m|$-values) outside the band. The pickup and kicker-behaviour is included via their impedances, detailed in Chap. 3.

Finally we remark that Eq. (5.22) can be written in various other forms involving e.g. sums over positive m only which clearly reveal the $(n \pm q)$ bands.

In the remainder of this sub-section we recollect terms to express $G(\omega_m)$, in general and with special reference to loop type pickups and kickers. We have

$$G(\omega_m) = \underbrace{e f_{rev} Z'_{PU\perp}}_{\substack{\text{`Voltage'}\\ \text{\&pickup}}} \cdot \underbrace{\lambda_a \cdot K_{KK\perp}}_{\substack{\text{Amplifier}\\ \text{\&Kicker}}} \cdot \underbrace{e^{im\omega_{rev}(t_c - t_{PK})}}_{\substack{\text{Delay}\\ \text{signal -particle}}} \cdot \underbrace{2\pi\, Q f_{rev}^2 \beta_K^{1/2} \beta_{PU}^{1/2} \frac{e}{m_0 c^2 \gamma\beta^2}}_{\text{Beam dynamics}}$$

$$(5.26)$$

where:

$Z'_{PU\perp}(\omega_m)$ is the total transverse impedance of the pickup system (composed of the impedances $Z'_{P\perp} = Z_{P\perp}/x$ (Table 3.3) of the n_{PU} units),

$K_{KK\perp}(\omega_m)$ the total voltage transfer function of the kicker system (composed of the transfer functions $K_\perp$ (Table 3.3) of the n_K kicker units),

$\lambda_a(\omega_m)$ the (voltage) transfer of the amplifier and signal path.

All three transfer functions are complex functions of ω_m [Eq. (5.23)] and introduce additional phase-shifts. Usually one aims at a design of the signal transmission system that brings the overall phase-shift close to zero for all ω_m within the band, with $\Delta\psi$ chosen according to Eq. (5.24).

We recall the pickup and kicker transfer functions for the special case of loop couplers. We assume n_{PU} pickup and n_K kicker-units. The power of each unit is coupled out and added with the proper delay in a combiner. It is then amplified and distributed onto the kicker units via splitters. We assume for the moment small and centred beam so that $Z'_{P\perp}$ and $K_\perp$ are independent of x. Then (looking at vertical cooling):

$$Z'_{PU\perp}(\omega_m) = \sqrt{n_{PU}}\, Z'_{P\perp}$$

$$\approx \sqrt{n_{PU}} \cdot \sqrt{2 R_P Z_{0,PU}} \cdot \tanh\left(\frac{\pi w}{2h}\right) \frac{1}{h} \cdot e^{i(\pi/2 - \Theta)} \sin(\Theta)$$

$$K_\perp(\omega_m) = \sqrt{n_K}\, K_\perp \approx \sqrt{n_K} \cdot \sqrt{\frac{Z_{0,K}}{2 R_K}} \cdot \tanh\left(\frac{\pi w}{2h}\right) \frac{2\ell}{h}\left(1 + \frac{v_{beam}}{v_{line}}\right) \cdot e^{-i\Theta} \frac{\sin(\Theta)}{\Theta}$$

(see Table 3.3, Chap. 3). For the second type of pickup and kicker (Fig. 3.2) we can establish similar relations, in this case for the horizontal plane, with the substitutions $Z'_{p\perp} \to Z'_{\equiv\equiv}$ and $K_\perp \to K_{\equiv\equiv}$ (Table 3.3, Chap. 3).

For a large beam, the pickup and kicker geometry functions deviate from linear change and constant behaviour respectively over the beam size (see Figs. 3.5 and 3.6 in Chap. 3). This can be especially pronounced for the second type of pickup and kicker (Fig. 3.2), where the beam will often be wider than the height of the device. In this case Eq. (5.17) becomes nonlinear. It can be solved approximately e.g. by the method of "harmonic balance" [42]: Assume e.g. that due to the nonlinearity of $Z'_{\equiv\equiv}$ and $K_{\equiv\equiv}$ the r.h.s. of (5.20) becomes

$$x_i e^{-i\Delta\psi} \cdot \sum_{m=-\infty}^{\infty} G(\omega_m) \cdot e^{i\varphi_m} \to x_i e^{-i\Delta\psi} \cdot \left(1 + \alpha_2 x_i^2 + \alpha_4 x_i^4 \ldots\right)$$

$$\cdot \sum_{m=-\infty}^{\infty} G_0(\omega_m) \cdot e^{i\varphi_m} \tag{5.27}$$

Insert a trial solution $x_i = \tilde{x}_i \cdot \sin(\omega_\beta t)$ and take only the linear terms ($x_i^3 \approx \frac{3}{4}\tilde{x}_i^3 \cdot \sin(\omega_\beta t) \cdots$) in the development of $\sin^3, \sin^5$ etc. In this way the nonlinearity is converted to a dependence on amplitude ($\tilde{x}_i$). For the damping rate one obtains

$$1/\tau \approx (1/\tau_0)\left(1 + \frac{3}{4}\alpha_2\tilde{x}_i^2 + \frac{5}{8}\alpha_4\tilde{x}_i^4 \ldots\right) \tag{5.28}$$

Clearly $1/\tau_0$ is the small amplitude rate calculated before (Eq. (5.22)).

The amplitude dependence can thus be determined from the pickup and kicker geometry. It can be averaged to obtain the "cooling rate of the beam".

Having established the interaction of the particle with itself we next include the noise due to the other particles and the electronic system.

5.4 Noise

Noise is treated in detail in Refs. [43–45]. For convenience we repeat the essentials here, which are useful to include the incoherent effect in the frequency-domain analysis of stochastic cooling.

Look at an oscilloscope picture like Fig. 5.3 which displays a pickup signal when 'no beam is in the machine', i.e. the electronic noise of the system. It is customary to represent the mean square (averaged over a long enough time T) of such noisy voltages by a pseudo Fourier transformation

$$\overline{u^2(t)} = \int_{-\infty}^{\infty} \phi(\omega)d\omega \tag{5.29}$$

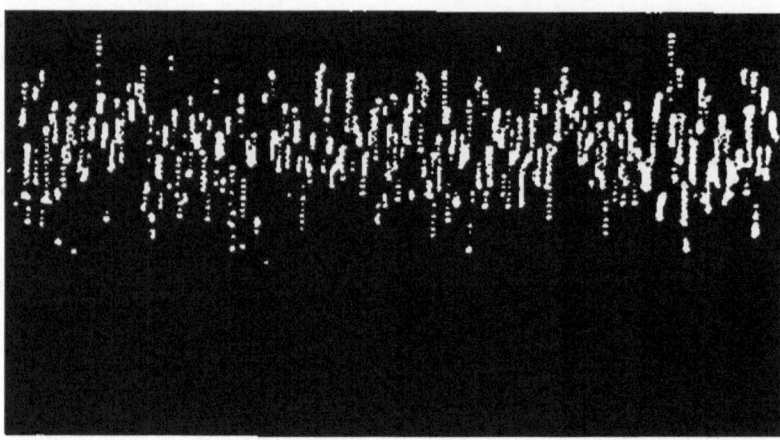

Fig. 5.3 Noise signal on a pickup (By courtesy of CERN, (©) CERN)

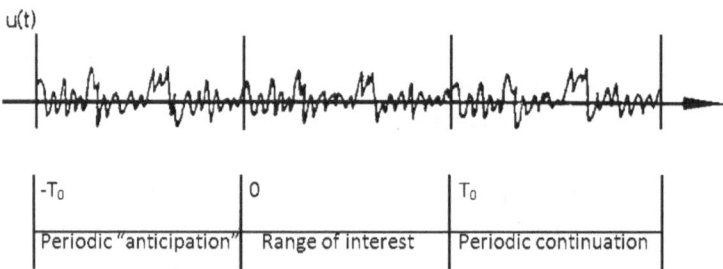

Fig. 5.4 A noisy voltage $u(t)$ observed from time $t = 0$ to $t = T_0$ during a long time interval: the function and its periodic continuation outside this range to permit a Fourier development (By courtesy of CERN, (©) CERN)

The 'spectral power density function $\phi(\omega)$' is closely related to a Fourier development of $u(t)$. In all practical applications the noisy voltage has been 'switched on' at some time $t = 0$ and we observe it for a long time-span up to $t = T_0$. Outside this range the waveform is irrelevant, so, for the purpose of computation we can periodically continue it (Fig. 5.4). We then deal with a periodic function $u(t \pm nT_0) = u(t)$, which we can Fourier-expand in the usual way

$$u(t) = \sum_{-\infty}^{\infty} \tilde{u}_m e^{im\omega_0 t}; \quad \omega_0 = 2\pi / T_0 \qquad (5.30)$$

The Fourier amplitudes

$$\tilde{u}_m = \frac{1}{T_0} \int_0^{T_0} u(t) e^{-im\omega_0 t} \, dt \qquad (5.31)$$

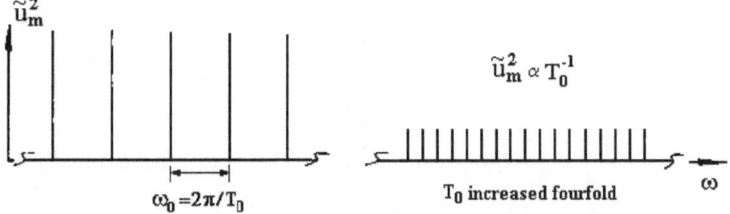

Fig. 5.5 The power spectrum $|u(\omega)|^2$ of a noisy voltage which is observed for a time interval T_0 and periodically continued outside this interval. Increasing T_0 the height of the spectral lines decrease proportionally to their spacing ω_0 so that the quantity $\tilde{u}_m^2/\omega_0$ remains the same. In the limit $T \to \infty$ one has a continuous spectrum where $u^2(\omega)/d\omega \to \phi(\omega)$ is the spectral power density function of Eq. (5.29) (By courtesy of CERN, (©) CERN)

are in general complex but for real $u(t)$, $\tilde{u}_{-m}$ is the conjugate of $\tilde{u}_m$. The mean square of Eq. (5.30) over the observation time T_0 is by definition

$$\overline{u^2(t)} = \frac{1}{T_0} \int_0^{T_0} u^2(t)dt$$

This yields after some calculation (transforming the square of the sum into a double sum, similar to the analysis in conjunction with Eq. (5.17), and noting that averaged over a period all $e^{ik\omega_0 t}$- terms vanish except for $k = 0$)

$$\overline{u^2(t)} = \sum_{m=-\infty}^{\infty} \tilde{u}_m \cdot \tilde{u}_{-m} = \sum_{m=-\infty}^{\infty} |\tilde{u}_m|^2 \tag{5.32}$$

This is known as Parseval's equation in the theory of Fourier series; it applies to any Fourier development! Equation (5.35) presents the 'average noise power' $u^2(t)$ as the sum of its spectral contributions at frequencies $\omega = m2\pi/T_0$. Analysed over shorter or longer time T_0 the spectra are as sketched in Fig. 5.5.

As the sum of the rays ('height x number') is $\overline{u^2(t)}$ in both cases, their height scales proportionally to their spacing. For very large observation time ($T_0 \to \infty$) the spectrum is continuous and the sum Eq. (5.35) is approximated by an integral:

$$\overline{u^2(t)} = \sum_{m=-\infty}^{\infty} |\tilde{u}_m|^2 = \sum_{m=-\infty}^{\infty} \frac{|\tilde{u}_m|^2}{\Delta\omega} \Delta\omega \to \int_{-\infty}^{\infty} \frac{|\tilde{u}_m|^2}{d\omega} d\omega \tag{5.33}$$

Hence we identify (for $T \to \infty$)

$$\frac{|\tilde{u}(\omega)|^2}{d\omega} \to \phi(\omega) \tag{5.34}$$

This interpretation permits us to calculate $\phi(\omega)$ and to establish the following theorem:

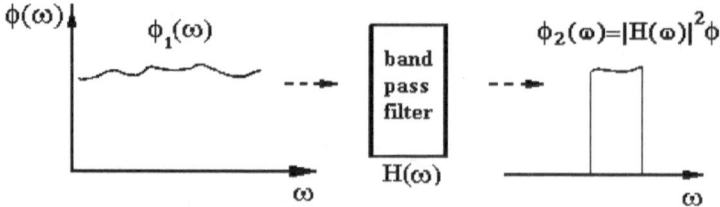

Fig. 5.6 An example of the theorem [Eq. (5.35)]: broadband noise transmitted through an ideal band-pass filter is converted into a band-limited noise (By courtesy of CERN, (©) CERN)

When noise with a spectral power density $\phi_1(\omega)$ is transmitted through a linear system with a (complex) transfer function $H(\omega)$ then the power spectrum at the output is

$$\phi_2(\omega) = \left|H(\omega)\right|^2 \phi_1(\omega) \tag{5.35}$$

This follows immediately from the preceding noting that each of the components on the r.h.s. of Eq. (5.30) when transmitted through the network transforms according to:

$$\tilde{u}_{m2} = H(m\omega_0) \cdot \tilde{u}_{m1}.$$

An example of the theorem (5.35) is the transformation of 'broadband noise' into band limited noise by a band-pass filter (Fig. 5.6).

This is as much as we need about noise for the purpose of this chapter.

5.5 Beam Response to a Noisy Kicker

Consider a linear oscillator driven by a noisy excitation $u(t)$ with power density $\phi(\omega)$.

$$\ddot{x} + \omega_\beta^2 x = u(t) \tag{5.36}$$

This problem was treated (in a more general context) in a classical paper by Hereward and Johnsen [45]. The result, the 'Hereward–Johnsen theorem', may—in our present case—be stated as follows:

For a particle injected at $t = 0$, the square of the amplitude of x, expected at time t is

$$\tilde{x}^2 = \frac{2\pi}{\omega_\beta^2} \cdot \phi(\omega_\beta) \cdot t \tag{5.37}$$

In words: the amplitude grows in a diffusion-like manner [$(\tilde{x}(t) \propto \sqrt{t}$, Fig. 5.7], at a rate which is determined by the spectral density of the noise *at the resonance frequency* ω_β.

Equation (5.37) is for a simple harmonic oscillator. If we inject noise into the cooling loop (or directly onto the kicker) we have again to include the 'sampling

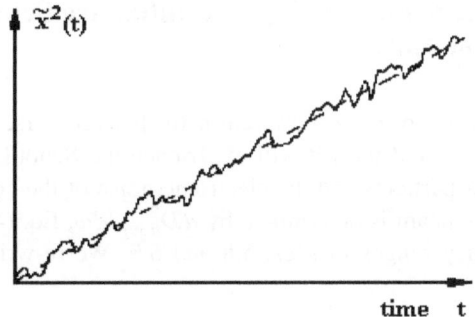

Fig. 5.7 Amplitude of betatron oscillation of a particle driven by a noisy kicker. The expectation value of x^2 grows linearly in time at a rate given by the spectral density of the noise at the resonance frequency. In addition to this average growth there is a fluctuating motion which is of little importance for the long-term behaviour (By courtesy of CERN, (©) CERN)

factor' $\sum e^{i\ell\omega_{rev}t}$ of Eq. (5.14) because we use a short localised kicker. Thus we write

$$\ddot{x} + \omega_\beta^2 x = u(t) \cdot \sum_{\ell=-\infty}^{\infty} e^{i\ell\omega_{rev}t} \tag{5.38}$$

The effect of each component, $e^{i\ell\omega_{rev}t}$, is to 'shift' the frequency content of the driving force $\omega \to \omega + \ell\omega_{rev}$. In this sense we may interpret the r.h.s. of Eq. (5.38) as a sum of noisy driving forces with frequencies

$$\omega_d \to \omega + \ell\omega_{rev}.$$

We can apply the Hereward-Johnsen result to each of these bands. Hence the response of Eq. (3.24) to a noise $u(t)$ is:

$$\tilde{x}^2 = \frac{2\pi}{\omega_\beta^2} \cdot t \cdot \sum_{\ell=-\infty}^{\infty} \phi(\omega_\beta + \ell\omega_{rev}) \tag{5.39}$$

Equivalently if you prefer to work with positive frequencies only, you may write Eq. (5.39) as

$$\tilde{x}^2 = \frac{2\pi}{\omega_\beta^2} \cdot t \cdot \left(\sum_{\ell=0}^{\infty} \phi((\ell+q)\omega_{rev}) + \sum_{\ell=1}^{\infty} \phi((\ell-q)\omega_{rev}) \right) \tag{5.40}$$

In working out the long term average of $\tilde{x}^2$ leading to Eqs. (5.39) and (5.40) we have used the fact that cross terms between different bands average to zero and also that $\phi(\omega) = \phi(-\omega)$.

Equations (5.39) and (5.40) clearly present the amplitude growth in terms of the spectral density of the noise at the betatron sidebands.

5.6 Back to Transverse Cooling, the Influence Schottky Noise and Amplifier Noise

After the digression to noise we now return to the transverse cooling of a test-particle. Apart from its self-term it will experience the Schottky noise due to the presence of the other particles and the electronic noise of the cooling system. The Schottky noise of the beam is determined by $dD_{n\pm q}^2/d\omega$, Eq. (4.16), and depicted for different frequency ranges in Figs. 5.8 and 5.9. We re-write Eqs. (4.19) and (4.20) as

$$\frac{dD_{n\pm q}^2}{d\omega} = M_{n\pm q}\phi_1$$

with $\phi_1 = N\frac{e^2 f_{rev}}{2\pi}\tilde{x}_{rms}^2$ the noise density Eq. (4.20) for band overlap and

$$M_{n\pm q} = \begin{cases} \frac{dN}{d\omega}\frac{\omega_{rev}}{2N} \approx \frac{\omega_{rev}}{2n\Delta\omega_{rev}} & \text{for separate bands} \\ \frac{dN}{d\omega}\frac{\omega_{rev}}{N} \approx \frac{\omega_{rev}}{n\Delta\omega_{rev}} & \text{for partial overlap} \\ 1 & \text{for complete overlap} \end{cases} \tag{5.41}$$

(where $\Delta\omega_{rev}$ is the total width of a band a).

The dimensionless quantity $M_{n\pm q}$ is the mixing factor which for longitudinal cooling was already introduced in Eq. (4.26).

We now turn to the electronic noise which referred to the entrance of the amplifier (exit of the pickup) has the spectral density Eq. (4.32). Hence we have to establish the noise to signal ratio $U_{n\pm q}$ at the amplifier entrance. This was done already in Sect. 4.4 with the result of Eq. (4.36). The total noise power density send into the system may then be written as

$$\phi(\omega) = \phi_1(\omega)\left[M_{n\pm q}(\omega) + U_{n\pm q}(\omega)\right] = N\frac{e^2 f_{rev}}{2\pi}\tilde{x}_{rms}^2\left[M_{n\pm q}(\omega) + U_{n\pm q}(\omega)\right]$$
$$\tag{5.42}$$

The noise is transmitted through the cooling loop (beam to kicker) in the same way as the dipole current signal of the test-particle. Hence the transfer function $(G(\omega_m)/(ef_{rev})$, see the discussion preceding Eq. (5.17)) is the same except for different phase factors. As far as the noise from the coasting beam is concerned, these phases are irrelevant and we have only to worry about the modulus of the transfer function. The spectral power density as experienced by the beam is therefore

$$\phi_k(\omega) = N\frac{|G(\omega)|^2}{4\pi f_{rev}}\tilde{x}_{rms}^2\left[M_{n\pm q}(\omega) + U_{n\pm q}(\omega)\right] \tag{5.43}$$

To establish the reaction of the test particle we have to add up the response at all the sideband frequencies $\omega_{n\pm q} = (n\pm q)\omega_{rev}$ in the pass-band. Using the Hereward-Johnson result, Eq. (5.40), we obtain the expectation value for the amplitude of the

Fig. 5.8 Transverse Schottky noise (rectangular approximation) for separated bands (low frequency), partial overlap (intermediate frequency) and complete band overlap (By courtesy of CERN, (©) CERN)

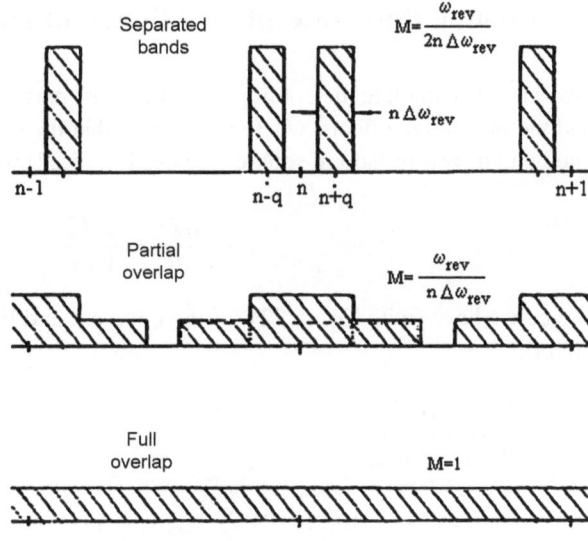

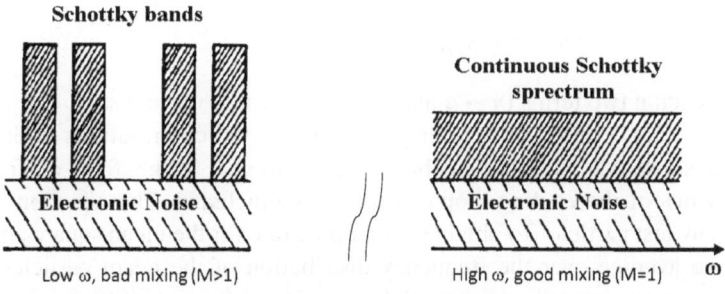

Fig. 5.9 The noise seen by a test particle is the Schottky noise due to the other particles and the electronic noise of the pre-amplifier and other low level components. At low frequency the Schottky noise occurs in bands with a density M times higher than in the situation of complete overlap. This increase of noise density corresponds to enhanced 'heating of the test particle due to bad mixing' (By courtesy of CERN, (©) CERN)

test particle as

$$\tilde{x}_i^2 = t \cdot \frac{N\tilde{x}_{rms}^2}{2\omega_\beta^2 f_{rev}} \sum_{n=n_{\min}}^{n_{\max}} \left\{ \left| G(\omega_{n+q}) \right|^2 \left[M(\omega_{n+q}) + U(\omega_{n+q}) \right] \right.$$

$$\left. + \left| G(\omega_{n-q}) \right|^2 \left[M(\omega_{n-q}) + U(\omega_{n-q}) \right] \right\} \tag{5.44}$$

Equation (5.44) determines 'the incoherent effect'.

5.7 Cooling Rate, Interplay of Coherent and Incoherent Effects

We can now calculate the cooling rate of the test particle by adding up the coherent and the incoherent effects, where we assume that the two contributions are uncorrelated and therefore 'add in square'. To use Eq. (5.22) we note that in general

$$\frac{1}{\tilde{x}^2}\frac{d\tilde{x}^2}{dt} = \frac{2}{\tilde{x}}\frac{d\tilde{x}}{dt}$$

Hence we have (with Eq. (5.22) for the coherent and Eq. (5.44), for the incoherent effect)

$$\left.\frac{d\tilde{x}^2}{dt}\right|_c + \left.\frac{d\tilde{x}^2}{dt}\right|_{ic}$$

$$= -\tilde{x}^2 \cdot \frac{1}{\omega_\beta} \mathrm{Im}\left[e^{-i\Delta\psi}\sum_{n_{\min}}^{n_{\max}} G(\omega_{n\pm q})e^{i\varphi_m(\omega_{n\pm q})}\right]$$

$$+ \tilde{x}_{rms}^2 \cdot \frac{\pi}{\omega_\beta^2\omega_{rev}}N\sum_{n_{\min}}^{n_{\max}}\left|G^2(\omega_{n\pm q})\right|\left[M(\omega_{n\pm q}) + U(\omega_{n\pm q})\right] \qquad (5.45)$$

where for each n two terms $(n-q$ and $n+q)$ enter into the sums.

Equation (5.45) represents cooling as a sum of the contributions at the $n+q$ and $n-q$ sidebands with two sidebands per harmonic. In this form all frequency characteristics of the cooling loop can readily be included. The equation is so far valid for any test particle. To obtain the damping rate for the mean square amplitude we have to average over the frequency distribution of the beam particles. As an approximation we can take a test-particle with $\tilde{x}^2 = \tilde{x}_{rms}^2$ as typical.

We re-discover imperfect mixing M on the way pickup to kicker. It is expressed here as enhancement of the heating by Schottky noise which is concentrated into bands and hence increased in density. Good mixing $(M = 1)$ corresponds in this picture to overlap of Schottky bands.

In the simple (but unrealistic!) case where $\varphi_m = 0$, $x = x_{rms}$ and $M(\omega) = M$, $G(\omega) = G$, $U(\omega) = U$ are constant in the pass-band, one recovers the familiar result [Eq. (2.42)]

$$\frac{1}{\tau_{x^2}} = -\frac{1}{x_{rms}^2}\frac{dx^2}{dt} = \frac{W}{N}\left[2g\sin\Delta\psi - g^2(M+U)\right]$$

$$\text{by calling } \frac{NG}{\omega_\beta f_{rev}} \equiv g \qquad (5.46)$$

One can express the sums in Eq. (5.45) by averages over the pass-band. With this interpretation one may write this equation in various different forms useful for comparison with previous results, for instance:

$$\frac{d\tilde{x}^2}{dt} = \frac{W}{N}\{-2\tilde{x}^2\langle \mathrm{Re}[g_{(\omega)}e^{i(\varphi_{m(\omega)}-\delta\Delta\psi)}]\rangle_{passband}$$

$$+\tilde{x}_{rms}^2\langle|g^2(\omega)|[M_{(\omega)}+U_{(\omega)}]\rangle_{passband}\} \qquad (5.47)$$

where

$$\delta\Delta\psi = \Delta\psi - \frac{\pi}{2}$$

$$g_{(\omega_{n\pm q})} = \frac{NG_{(\omega_{n\pm q})}}{\omega_\beta f_{rev}}$$

Further the average over the $2(n_{max} - n_{min}) = 2W/f_{rev}$ betatron sidebands enters, which is explicitly

$$\langle \mathrm{Re}[g_{(\omega)}e^{i(\varphi_{m(\omega)}-\delta\Delta\psi)}]\rangle_{passband}$$

$$=\frac{f_{rev}}{2W}\sum_{n_{min}}^{n_{max}}(\mathrm{Re}\{g_{(\omega_{n+q})}e^{i(\varphi_{m(\omega_{n+q})}-\delta\Delta\psi)}\}+\mathrm{Re}\{g_{(\omega_{n-q})}e^{i(\varphi_{m(\omega_{n-q})}-\delta\Delta\psi)}\})$$

$$\langle|g_{(\omega)}^2|[M_{(\omega)}+U_{(\omega)}]\rangle_{passband}$$

$$=\frac{f_{rev}}{2W}\sum_{n_{min}}^{n_{max}}(|g_{(\omega_{n+q})}^2|[M_{(\omega_{n+q})}+U_{(\omega_{n+q})}]+|g_{(\omega_{n-q})}^2|[M_{(\omega_{n-q})}+U_{(\omega_{n-q})}])$$

Equation (5.47) expresses cooling by an average over the pass band and accounts for

- gain variation by $g_{(\omega)}$,
- the error in the betatron phase advance pickup to kicker by $\delta\Delta\psi$,
- the unwanted mixing and other phase errors by $\varphi_{m(\omega)}$,
- the imperfection of the wanted mixing by $M_{(\omega)}$,
- the electronic noise by $U_{(\omega)}$.

This concludes our detailed excursion to transverse cooling. The power needed is discussed in Appendix C. In Chap. 7 we will briefly sketch the treatment of transverse cooling by the Fokker-Planck equation.

Chapter 6
Feedback via the Beam-Signal Shielding

6.1 Transverse Motion

We shall now attempt to introduce a further ingredient of cooling theory known as 'feedback via the beam' or 'signal shielding' [40]. Although this refinement will change our previous results by at most a factor of 2, the change of the beam signals when the cooling loop is closed/opened has become an important diagnostics tool [46, 47]. We will start with the effect on transverse cooling and then generalise to include the longitudinal case.

A question that comes to mind is: Where did we miss out this effect in our treatment so far? In fact considering the test-particle equation for betatron cooling we wrote

$$\ddot{x} + \omega_\beta^2 x = \underbrace{Gx}_{\substack{\text{effect of particle} \\ \text{upon} \\ \text{itself: coherent} \\ \text{term}}} + \underbrace{\sum Gx_i}_{\substack{\text{effect of other particles:} \\ \text{Schottky, noise,} \\ \text{fluctuating term with} \\ \text{zero average}}} + \underbrace{\text{"system noise"}}_{\substack{\text{electronic noise} \\ \text{fluctuating with zero} \\ \text{average} \\ \text{O.K.}}} \qquad (6.1)$$

We described the effect of the other particles, $\sum Gx_i$ as Schottky noise of an undisturbed beam, i.e. as a fluctuating term with zero time average. This assumption is not generally correct.

In fact the noise in a beam subject to cooling is different from the free beam noise. The feedback of the cooling signals via the beam changes all ingredients of the analysis, namely beam noise as well as the influence of the coherent term and the electronic noise, as illustrated by Fig. 6.1. Fortunately F. Sacherer has also shown the road to rescue our previous results [40].

As a pre-exercise: consider a system of N oscillators with a harmonic driving force and a 'collective force' proportional to the sum of the displacements of these oscillators. Take for the jth particle—sorry—oscillator

$$\ddot{x}_j + \omega_j^2 x_j = \underbrace{\tilde{V}e^{i\omega t}}_{\substack{\text{harmonic} \\ \text{driving force}}} + \underbrace{G \cdot N\langle x \rangle}_{\substack{\text{collective} \\ \text{force}}} \qquad (6.2)$$

D. Möhl, *Stochastic Cooling of Particle Beams*, Lecture Notes in Physics 866, DOI 10.1007/978-3-642-34979-9_6, © Springer-Verlag Berlin Heidelberg 2013

Fig. 6.1 Cooling system including the coherent beam modulation x_b imposed at the kicker and partially preserved up to the pickup due to imperfect mixing. The lower diagram shows Sacherer's equivalent feedback loop. Amplifier noise (x_n) and Schottky noise (x_s) are random noises whereas the coherent modulation is fed back via the beam from kicker to the pickup. This feedback changes the open loop response by a complex transfer function $T(\omega)$, which depends on the amplification and the degree of mixing between kicker and pickup (By courtesy of CERN, (©) CERN)

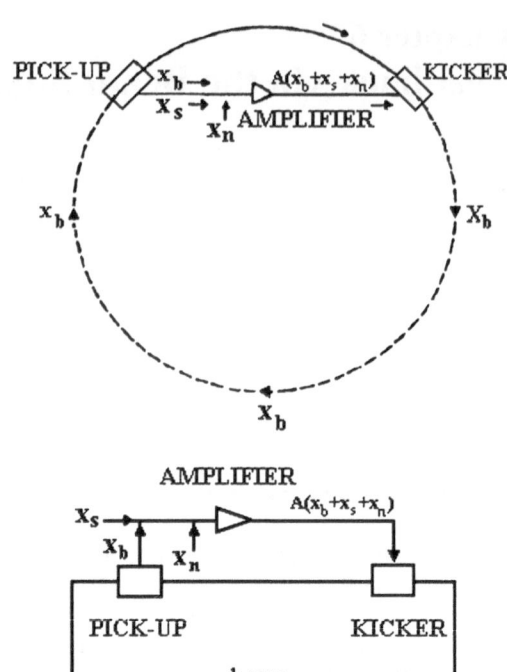

Here G is a constant giving the strength of the coupling, and $\langle x \rangle = \frac{1}{N} \sum_{k=1}^{N} x_k$ is the average beam displacement; $G \cdot N \langle x \rangle$ is total feedback from N particles.

To solve Eq. (6.2) we insert a trial solution

$$x_j = \tilde{x}_j e^{i\omega t}, \qquad \langle x \rangle = \langle \tilde{x} \rangle e^{i\omega t}$$

and find for the amplitude

$$\tilde{x}_j = \frac{1}{\omega_j^2 - \omega^2} \left[\tilde{V}(\omega) + G \cdot N \langle \tilde{x} \rangle \right] \tag{6.3}$$

Average and call:

$$G \cdot N \left\langle \frac{1}{\omega_j^2 - \omega^2} \right\rangle = S(\omega) \tag{6.4}$$

Solve for

$$\langle \tilde{x}_j \rangle = \frac{1}{N} \sum_{j=1}^{N} \tilde{x}_j = \langle \tilde{x} \rangle$$

to find

$$\langle \tilde{x} \rangle = \frac{S(\omega)}{1 - S(\omega)} \cdot \frac{\tilde{V}}{G \cdot N} \tag{6.5}$$

Fig. 6.2 A 'mechanical analogue' of Eq. (6.2). Person V tries to excite a system of oscillators (masses an springs) by shaking their point of suspension. Person G tries to damp the motion by observing the average displacement $\langle x \rangle$ of the oscillators and applying a damping force $G\langle x \rangle$ (By courtesy of CERN, (©) CERN)

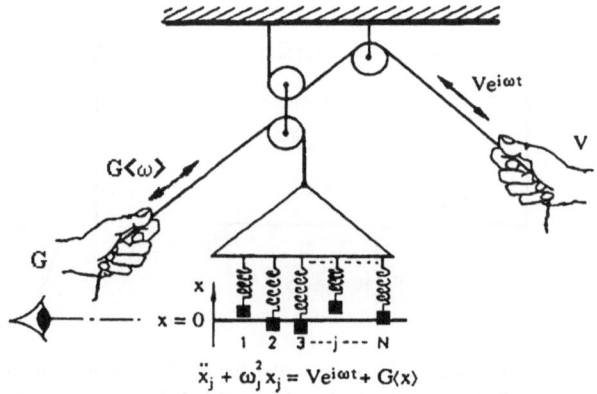

$$\ddot{x}_j + \omega_j^2 x_j = Ve^{i\omega t} + G\langle x \rangle$$

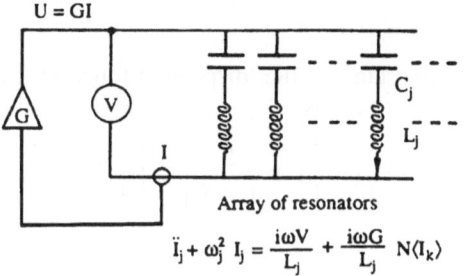

$$\ddot{I}_j + \omega_j^2 I_j = \frac{i\omega V}{L_j} + \frac{i\omega G}{L_j} N\langle I_k \rangle$$

Fig. 6.3 An electrical analogue of Eq. (6.2). A group of LC-resonators is driven by a voltage $V = e^{i\omega t}$. The sum $I = \sum I_k = N\langle I_k \rangle$ of the currents through the resonators is fed back through an amplifier with gain G to add an input voltage $G \cdot \sum I_k$ proportional to the sum of the currents (By courtesy of CERN, (©) CERN)

Thus we do have a finite coherent amplitude ($\langle \tilde{x} \rangle$). We can now use Eq. (6.5) to eliminate the 'collective force' term from Eq. (6.2). We find

$$\ddot{x}_j + \omega_j^2 x_j = \tilde{V}(\omega)e^{i\omega t} \underbrace{\left[1 + \frac{S(\omega)}{1 - S(\omega)}\right]}_{\text{shielding factor}} \qquad (6.6)$$

A 'mechanical' and an electrical analogue of Eq. (6.2) are sketched in Figs. 6.2 and 6.3.

Instead of treating the original Eq. (6.2) with

$$\text{r.h.s.} = [\text{driving force}] + [\text{weighted coherent displacement}]$$

we can therefore treat the same equation with the more convenient:

$$\text{r.h.s.} = [\text{driving force}] \times [\text{shielding factor}].$$

This is the essence of Sacherer's 'trick'.

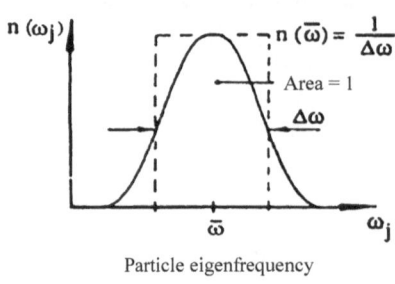
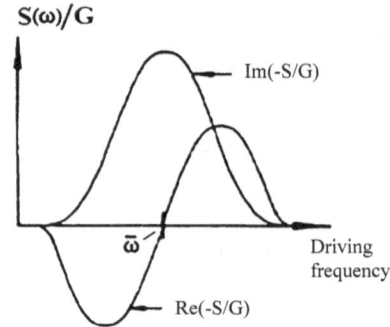

Fig. 6.4 Frequency distribution and typical behaviour of the dispersion function Eq. (6.8) for a given function $G(\omega)$ which is constant (or slowly varying) near the beam response frequency $\bar{\omega} = \omega_\beta$ (By courtesy of CERN, (©) CERN)

A quantity of key importance is the 'dispersion function' $S(\omega)$ entering into the shielding factor

$$T(\omega) = \left[1 + \frac{S}{1-S}\right] = \left[\frac{1}{1-S}\right] \tag{6.7}$$

For large N we have

$$S(\omega) = G \cdot N \left\langle \frac{1}{\omega_j^2 - \omega^2} \right\rangle = G \sum \frac{1}{\omega_j^2 - \omega^2}$$

$$= N \cdot G \int \frac{n(\omega_j)}{\omega_j^2 - \omega^2} d\omega_j \approx \frac{N \cdot G}{2\omega_\beta} \int \frac{n(\omega_j)}{\omega_j - \omega} d\omega_j \tag{6.8}$$

Here $n(\omega_j)d\omega_j$ is the fraction of particles with eigenfrequencies in a band of width $d\omega_j$ near ω_j and $\int n(\omega_j)d\omega_j = 1$.

Dispersion integrals of the type (6.8) are treated in the theory of Landau damping of coherent beam instabilities (see e.g. [48–50]). For convenience some features are repeated in Appendix D. Due to the pole in the integrand, $S(\omega)$ has an imaginary part even if G real. Details depend on the distribution $n(\omega_j)$ of eigenfrequencies. A typical behaviour of $S(\omega)$ is sketched in Fig. 6.4.

A an approximation it is useful to assume a 'semicircular' distribution

$$n(x) = \begin{cases} \frac{2}{\pi \Delta^2} \sqrt{\Delta^2 - (\omega_j - \omega_\beta)^2} & \text{for } |\omega_J - \omega_\beta| \leq \Delta = \Delta\omega/2 \\ 0 & \text{elsewhere} \end{cases}$$

For this one obtains (Appendix D)

$$-S(\omega) \approx \frac{N \cdot G}{\omega_\beta \Delta^2} \left[i \cdot \sqrt{\Delta^2 - (\omega - \omega_\beta)^2} + (\omega - \omega_\beta) \right] \tag{6.9}$$

To go one step further we now analyse the problem of beam excitation by a single harmonic driving force injected onto a cooling system when the loop is closed. The term neglected so far in betatron cooling is of the form $\sum_k G(\omega)x_k$, representing the linear influence of all N particles on the test particle. In fact omitting this term we have assumed that heating by the other particles proceeds only via the Schottky noise (see the analysis in Chap. 4). We shall now examine the influence of just this 'omitted contribution' when a strong driving force is injected.

In writing down the betatron equation we take account, once again, of localised pickup and kicker. For particle j we write

$$\ddot{x}_j + \omega_\beta^2 x_j = \left[\sum_k G(\omega) \sum_{m=-\infty}^{\infty} e^{im\omega_{rev}(t-t_k+t_j)-i\mu} x_k + \tilde{V}e^{i\omega t} \right]$$

$$\times \sum_{\ell=-\infty}^{\infty} e^{-i\ell\omega_{rev}(t-t_j-t_{PK})} \tag{6.10}$$

Here the first sum (k) is over the N beam particles, the sum over m is the 'sampling term' due to the localised pickup and the sum over ℓ represents the harmonics of the localised kick; t_k is the arrival time of particle k at the pickup, $x_k \rightarrow x_k e^{-i\mu}$ presents the transformation of its oscillation from pickup to kicker, t_d is the signal delay of the cooling loop, t_{PK} is the travelling time of particle j from pickup, where it arrives at t_j, to kicker, where its arrival time is $t_j + t_{PK}$; $\tilde{V}e^{i\omega t}$ is the external driving force.

Once again we drop all rapidly varying terms, i.e. we only retain harmonics with $m = -\ell$ in the sum over m in Eq. (6.10). In the sum over ℓ we only retain frequencies $\omega \pm \ell\omega_{rev} \approx \omega_\beta$ close to a betatron resonance. We assume that all bands are well separated so that only one ℓ leads to resonance. Thus we simplify Eq. (6.10) to

$$\ddot{x}_j + \omega_\beta^2 x_j = \left[\sum_k G(\omega)e^{i\ell\omega_{rev}(t-t_k+t_j)-i\mu} x_k + \tilde{V}e^{i\omega t} \right] e^{-i\ell\omega_{rev}(t-t_j-t_{PK})} \tag{6.11}$$

As response to the driving term $\tilde{V}e^{i\omega t}$ we expect a solution of the form

$$x_j = \tilde{x}_j e^{i(\omega-\ell\omega_j)t + i\ell\omega_{rev}(t_j+t_{PK})}$$

for any particle j. We substitute the corresponding expression for x_k on the r.h.s. of Eq. (6.11)] and continue in a similar fashion as we proceeded from Eqs. (6.2), (6.3) to Eq. (6.7) before. We find:

$$\tilde{x}_j = \frac{1}{\omega_\beta^2 - (\omega-\ell\omega_{rev})^2}\left(G \cdot N\langle\tilde{x}\rangle + \tilde{V}\right) \tag{6.12}$$

defining

$$\langle\tilde{x}\rangle = \frac{1}{N}\sum_{k=1}^{N} \tilde{x}_k e^{-i(\mu+\varphi_k)}$$

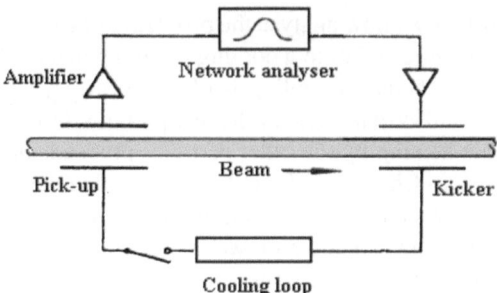

Fig. 6.5 Arrangement to measure beam transfer function. The frequency sweep of the network analyser is set to cover one or several betatron sidebands. The difference in beam response with cooling loop open and closed can be used to optimise the loop gain (By courtesy of CERN, (©) CERN)

Here $\varphi_k = \ell\omega_{rev}(t_k - t_{PK})$ is the synchronization error of particle k (with all phase shifts of the cooling loop included). We can now define a shielding factor for the present situation

$$T(\omega_\ell) = \left[\frac{1}{1 - S(\omega_\ell)}\right] \quad \text{with}$$

$$S(\omega_\ell) = \frac{N \cdot |G(\omega_\ell)|e^{-i\mu}e^{i\varphi_\ell}}{2\omega_\beta} \int \frac{n(\omega_\ell)}{\omega_\ell - \omega}d\omega_\ell \tag{6.13}$$

Note that $S(\omega_\ell)$ now contains the complex factors due to the betatron advance pickup to kicker and synchronization errors. Using shielding factor (6.13) we can rewrite Eq. (6.11) as

$$\ddot{x}_j + \omega_\beta^2 x_j = T(\omega_\ell)\tilde{V}e^{i\omega t} \tag{6.14}$$

Thus when the cooling loop is closed, the response of Eq. (6.10) to $\tilde{V}e^{i\omega t}$ (with $\omega \approx \omega_\ell$) changes by $T(\omega)$. In this way $T(\omega)$ can be observed and $G(\omega)$ can be deduced from it. Usually these measurements are done using a network analyser to display the beam response to a swept sine wave (beam transfer function measurement) as sketched in Fig. 6.5. This permits one to adjust the characteristics of the cooling loop band by band.

To complete our analysis we return to Eq. (6.10) but now assume a general driving force represented by a Fourier series (or a Fourier integral) with a spectral density function $V(\omega)$. We invoke superposition and resonant behaviour of the betatron equation at the frequencies $\omega_\ell = (Q \pm \ell)\omega_{rev}$. Thus we rewrite Eq. (6.10) as

$$\ddot{x}_j + \omega_j^2 x_j = \sum_{\ell=-\infty}^{\infty} T(\omega_\ell) \cdot \tilde{V}(\omega_\ell)e^{i\omega_\ell t} \tag{6.15}$$

where the shielding factors are given by Eq. (6.13). This equation presents the effect as the sum shielding interaction at the betatron sidebands. Each band now has its

Fig. 6.6 Typical behaviour
of the shielding factor $T(\omega_\ell)$
near a betatron sideband
frequency of the beam (By
courtesy of CERN,
(©) CERN)

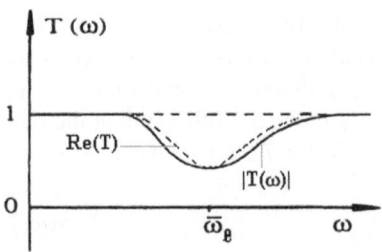

proper shielding factor (well separated bands, i.e. poor mixing assumed). The effect
of the shielding factor is fully equivalent to introducing a transfer function between
the driver and the kicker.

We now generalise the cooling rate, Eqs. (5.45), to include shielding. To this end
we interpret $\tilde{V}(\omega)$, Eq. (6.15), as the cooling signals discussed before (namely the
self-effect of the test particle, the Schottky noise due to the other particles and the
amplifier noise). Then we replace for each of the betatron bands involved the

r.h.s. = [coherent term] + [Schottky noise] + [amplifier noise]

+ [weighted beam displacement]

by r.h.s. = [coherent term + Schottky noise + amplifier noise] × [shielding factor].

Since the beam feedback acts like a transfer function we can simply include this
into the cooling equations of Chap. 5 [(like e.g. Eqs. (5.20) or (5.45)] by substituting
for each of the bands (ℓ)

$$G(\omega_\ell) \to G(\omega_\ell) \cdot T(\omega_\ell) \qquad (6.16)$$

Hence in the general case one has to include the shielding in Eq. (5.45) as

$$\frac{d\tilde{x}^2}{dt} = -\tilde{x}^2 \cdot \frac{1}{\omega_\beta} \, \mathrm{Im}\left[e^{-i\mu} \sum_{n_{\min}}^{n_{\max}} \left|G(\omega_{n\pm q})\right| e^{i\varphi_{n\pm q}} \cdot T(\omega_{n\pm q}) \right]$$

$$+ \tilde{x}_{rms}^2 \cdot \frac{\pi}{\omega_\beta^2 \omega_{rev}} N \sum_{n_{\min}}^{n_{\max}} \left|G^2(\omega_{n\pm q})\right| \cdot \left|T(\omega_{n\pm q})\right|^2 \left[M(\omega_{n\pm q}) + U(\omega_{n\pm q})\right]$$

$$(6.17)$$

The phase factors in the coherent term appear a for second time in the calculation of $T(\omega_{n\pm q})$. A typical behaviour of the shielding function (6.13) for $\mu =$
$\pi/2$ and $\varphi_{n\pm q} = 0$ is sketched in Fig. 6.6. For small gain ($N \cdot G$ and hence S small)
the shielding factor is close to 1.

To gain further insight we look at particles near the centre of the distribution
($\omega_\ell = \ell \omega_{rev} \pm \omega_\beta$) and assume perfect betatron phase and perfect signal delay
pickup to kicker. Let us introduce the 'reduced gain'

$$g(\omega_\ell) = \frac{|G(\omega_\ell)| \cdot N}{\omega_\beta \omega_{rev}/2\pi} \qquad (6.18)$$

similar to Eq. (6.17). Further we recall the definition of the mixing factor for well separated bands. For the 'semi-circular' distribution it has the central value $M_\ell = \omega_{rev}/(\frac{\pi}{2}\Delta\omega_\ell)$ instead of $M_\ell = \omega_{rev}/(2\Delta\omega_\ell)$ [Eq. (5.41)] for rectangular bands, $\Delta\omega_\ell$ being the full width at the base. Using Eq. (6.9) for the dispersion integral and (6.13) for $T(\omega_\ell)$ we have

$$-i \cdot S(\omega_\ell) \approx \frac{2N}{\omega_\beta}\frac{|G(\omega_\ell)|}{\Delta\omega_\ell} = \frac{g_\ell M_\ell}{2}$$

$$T(\omega_\ell) \approx \frac{1}{1 + g_\ell M_\ell/2}$$

(6.19)

The cooling rate equation for any particle is obtained from the expressions of Chap. 5 by replacing $G(\omega_\ell) \rightarrow G(\omega_\ell) \cdot T(\omega_\ell)$. We obtain, summing over the $2 \cdot (n_{max} - n_{min})$ bands within the pass-band:

$$\frac{1}{\tau} = \frac{f_{rev}}{4N} \sum_{passband} \left[\frac{2g_{n\pm q}}{1 + g_{n\pm q}M_{n\pm q}/2} - \frac{g_{n\pm q}^2}{(1 + g_{n\pm q}M_{n\pm q}/2)^2}(M_{n\pm q} + U_{n\pm q}) \right]$$

(6.20)

Here we assume cooling of a particle with $\tilde{x}^2 = \tilde{x}_{rms}^2$ (which is not fully consistent with the assumption of small-amplitude made for the calculation of the shielding factor).

Equation (6.20) is formally the same as the previous cooling equation (5.44) if we replace

$$g_{n\pm q} \rightarrow \frac{g_{n\pm q}}{1 + g_{n\pm q}M_{n\pm q}/2}$$

(6.21)

Optimum cooling is obtained from Eq. (6.20) when for all bands

$$g_{n\pm q} = \frac{1}{M_{n\pm q}/2 + U_{n\pm q}}, \quad g_{n\pm q} \rightarrow \frac{2}{M_{n\pm q}}$$

(6.22)

The limiting case $(\rightarrow)$ is for negligible electronic noise. The optimum shielding factor and the optimum rate corresponding to Eq. (6.22) are:

$$T_{n\pm q} = \frac{1}{1 + M_{n\pm q}/(M_{n\pm q} + 2U_{n\pm q})}, \quad T_{n\pm q} \rightarrow \frac{1}{2}$$

(6.23)

$$\frac{1}{\tau} = \frac{f_{rev}}{4N} \sum_{passband} \frac{1}{M_{n\pm q} + U_{n\pm q}}$$

(6.24)

Thus in the situation of negligible electronic noise, optimum cooling is obtained when the gain (at all bands involved) leads to signal reduction by a factor of about 2. By comparing open and closed loop signals (either Schottky noise or driven-beam response) the gain can be optimised band by band. Note that for negligible electronic noise the optimum gain Eq. (6.22) is twice the optimum (c.f. Eqs. (2.28) and (2.42)) calculated without beam feedback. When the electronic noise predominates, then

Fig. 6.7 Reduction of a
transverse Schottky band
when the cooling loop is
closed. With negligible
electronic noise and well
separated bands optimum
gain of the cooling loop
corresponds to a signal
amplitude reduction by about
2 in the centre of the bands
(By courtesy of CERN,
(©) CERN)

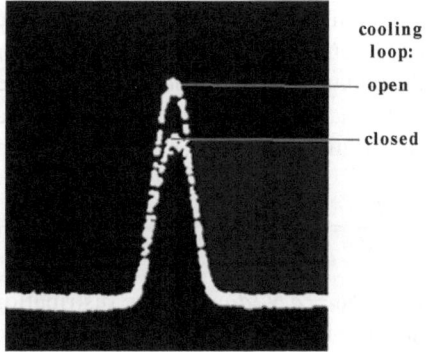

Eq. (6.22) yields the same optimum g as in the case without shielding. An example of Schottky signal shielding of a band is given in Fig. 6.7.

Thus the inclusion of beam shielding (which was done in an approximate manner here) leads to an improved expression for the cooling rate and—more importantly—to an adjustment criterion for the cooling system. The analysis done so far for betatron cooling will be repeated for momentum spread damping (next subsection) where similar gain adjustment criteria apply.

6.2 Longitudinal Beam Feedback

To determine the longitudinal beam feedback we first assume a beam with zero energy spread, excited by a small signal $Ve^{i\omega t}$ applied on a longitudinal kicker. This will modulate both the energy and the density of the beam. If ω is near a harmonic $n\omega_{rev}$ of the revolution frequency the modulation of the energy and the beam current I_0 are

$$\frac{E_1}{eV} \approx \frac{if_{rev}}{n\omega_{rev} - \omega} \tag{6.25}$$

$$\frac{I_1}{eV} = -\frac{if_{rev}n\kappa I_0}{(n\omega_{rev} - \omega)^2} \tag{6.26}$$

Here $\kappa = -\frac{n\omega_{rev}}{\beta^2 E}$, $\eta = \gamma_{transition}^{-2} - \gamma^{-2}$.

The derivation of Eqs. (6.25) and (6.26) ([19] and Appendix E) is similar to the treatment given by Hofmann [51] and Edwards and Syphers [52] in the context of longitudinal instability theory.

With an ideal sum pickup, as used for filter and time of flight cooling, we only see the current modulation, but for the more general case of a pickup with sensitivity $P(\omega_{rev})$ we find for the pickup current I_p

$$\frac{I_p}{eV} = P\frac{I}{V} + \frac{dP}{d\omega_{rev}}\kappa I_0 \frac{E_1}{eV} \tag{6.27}$$

For Palmer cooling the sensitivity function $P(\omega_{rev})$ is proportional to transverse deviation $x_p = D\Delta p/p = (-D/\eta)\Delta\omega_{rev}/\omega_{rev}$ where D is the orbit dispersion at the pickup. For filter and time of flight cooling $P(\omega_{rev}) = 1$.

For a beam with finite energy spread, we integrate Eq. (6.27) over ω_{rev} after substitution of Eqs. (6.25) and (6.26), taking $I_0 = ef_{rev}$ per particle:

$$\frac{I_p}{eV} = \frac{ief_{rev}^2\kappa}{n}\left[-\int\frac{\psi_0 P}{(\omega_{rev}-\omega/n)^2}d\omega_{rev} + \int\frac{\psi_0 \cdot dP/d\omega_{rev}}{(\omega_{rev}-\omega/n)}d\omega_{rev}\right] \quad (6.28)$$

Here $\psi_0 = dN/d\omega_{rev}$ is the distribution of particles with respect to revolution frequency. By partial integration of the first term, using the fact that $\psi_0 \to 0$ at the limits of the integration interval, we find

$$\frac{I_p}{eV} = -\frac{ief_{rev}^2\kappa}{n}\int\frac{P \cdot d\psi_0/d\omega_{rev}}{(\omega_{rev}-\omega/n)}d\omega_{rev} \quad (6.29)$$

Due to the resonant denominator the main contribution to the integral comes from frequencies $\omega_{rev} \approx \omega/n$. Hence we may approximate Eq. (6.29) by

$$\frac{I_p}{eV} \approx -\frac{ief_{rev}^2\kappa}{n} \cdot P(\omega/n)\int\frac{d\psi_0/d\omega_{rev}}{(\omega_{rev}-\omega/n)}d\omega_{rev} \quad (6.30)$$

provided that $P(\omega_{rev})$ does not change too rapidly in the neighbourhood of ω/n. The response (6.30) is equal to the response of a system with a sum pickup with $P = 1$, including a filter that reproduces the factor $P(\omega/n)$.

It is now somewhat more convenient to introduce the (total) energy E of a particle (in eV), instead of the revolution frequency, as the independent variable. We use the relation (see e.g. [3, 4]) between the fractional change of energy and revolution frequency

$$\frac{d\omega_{rev}}{\omega_{rev}} = -\frac{\eta}{\beta^2}\frac{dE}{E}, \quad \eta = \gamma_{transition}^{-2} - \gamma^{-2}$$

to redefine the density distribution

$$\psi = \frac{dN}{dE} = |\kappa|\frac{dN}{d\omega_{rev}} = |\kappa|\psi_0, \quad (6.31)$$

with κ given under Eq. (6.26). Then Eq. (6.30) becomes:

$$\frac{I_p}{eV} \approx -\frac{ief_{rev}^2}{n} \cdot P(E)\int\frac{1}{\kappa}\frac{d\psi/dE^*}{(E^*-E)}dE^* \quad (6.32)$$

Integrals of this form are well known in the theory of longitudinal instabilities [48–52]. Due to the pole at $E^* = E$ the integral in (6.29) has the residuum

$$i(\pi/|\kappa|)(d\psi/dE^*)_{E^*=E}$$

Fig. 6.8 Longitudinal beam response

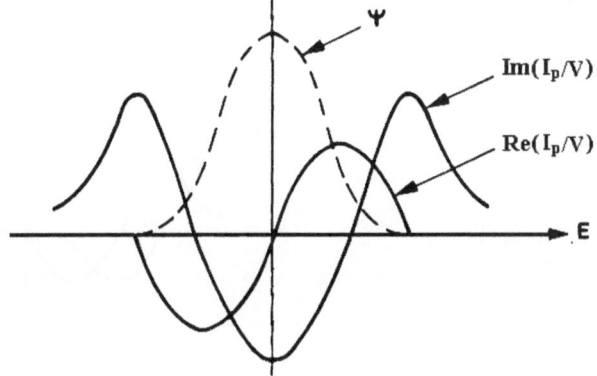

in addition to its principal value. The sign of the residuum depends on the nature of the problem and can be verified for the present case by assuming a small damping term ($\omega_{rev} \to \omega_{rev} + i\alpha$) and then proceeding to the limit $\alpha \to 0$, as it was done for transverse shielding in Appendix D.

Thus Eq. (6.29) finally can be written as

$$\frac{I_p}{eV} \approx -\frac{ef_{rev}^2}{n} \cdot P(E) \left[\frac{i}{\kappa} \int_{PV} \frac{d\psi/dE^*}{(E^* - E)} dE^* + \frac{\pi}{|\kappa|} \frac{d\psi}{dE} \right] \tag{6.33}$$

For a sum pickup, $P(E) = 1$, the typical variation of the two components with E is shown in Fig. 6.8 for a ring working above transition energy ($K < 0$); below transition, the imaginary component has the opposite sign.

The open-loop gain S_n at the nth harmonic for filter cooling follows from (6.33) by inserting $P(E) = 1$ and multiplying I_p/eV by $G'Z_{PU//}K_{KK//}$ where G' is the transfer function of the cooling loop from pickup to kicker and $Z_{PU//}$ and $K_{KK//}$ are the (total) pickup and kicker functions introduced in Chap. 3; G' includes the filter and the amplifier. The real and imaginary part of S depend on E roughly as shown in Fig. 6.9, again for a ring above transition. For Palmer cooling $P(E)$ depends on the transverse deviation $x_E = (D/\beta^2)(\Delta E/E)$. We insert $P(E) = (D/\beta^2)(\Delta E/E)Z'_{PU\perp}$ in Eq. (6.33) and multiply with $G'K_{KK//}$ to obtain S_n.

Thus we have

$$S_n \approx -\frac{ef_{rev}^2}{n} \cdot G'Z_{PU//}K_{k//} \left[\frac{i}{\kappa} \int_{PV} \frac{d\psi/dE^*}{(E^* - E)} dE^* + \frac{\pi}{|\kappa|} \frac{d\psi}{dE} \right] \tag{6.34}$$

for filter cooling and

$$S_n \approx -\frac{ef_{rev}^2}{n} \cdot (D/\beta^2)(\Delta E/E)Z'_{PU\perp}G'K_{k//} \left[\frac{i}{\kappa} \int_{PV} \frac{d\psi/dE^*}{(E^* - E)} dE^* + \frac{\pi}{|\kappa|} \frac{d\psi}{dE} \right] \tag{6.35}$$

for Palmer cooling.

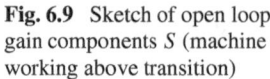

Fig. 6.9 Sketch of open loop gain components S (machine working above transition)

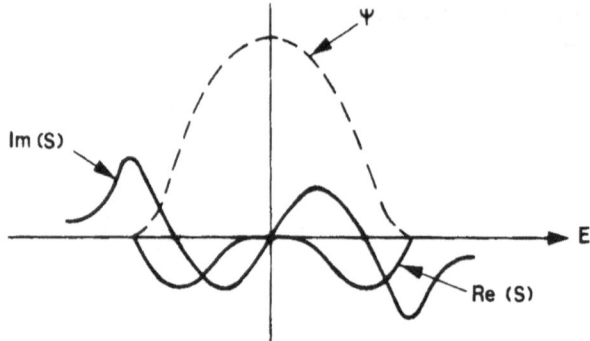

To include shielding in the longitudinal cooling we have to modify the nth harmonic of the coherent term $\text{Re}(\frac{1}{\tau_{c,n}}) \rightarrow \text{Re}(\frac{1}{\tau_{c,n}} \frac{1}{(1-S_n)})$ and the nth contribution the heating terms $(\frac{1}{\tau_{ic,n}}) \rightarrow (\frac{1}{\tau_{ic,n}} \frac{1}{|1-S_n|^2})$. This will be used in Chap. 7 in the Fokker-Planck equation.

Both components of the open-loop gain are zero in the centre. For most distributions (except those with steep gradients) their peak values are of the same order of magnitude. The imaginary part will tend to infinity at places where $d\psi/dE$ is discontinuous, e.g. at the edges of a parabolic distribution. This will locally affect the cooling in such a way that the discontinuity disappears; numerical calculations show that a slight rounding of the tails will strongly reduce the effect of $\text{Im}(S)$ on the cooling.

Although the beam response is quite different from the one found for the betatron case, because it depends on $d\psi/dE$ instead of ψ, the order of magnitude of the open-loop gain needed is still the same. The gain required will, however, depend on frequency in a complicated way that cannot be realised with practical filters; S will therefore vary across the distribution and we have to find a compromise value for the system gain.

Chapter 7
The Distribution Function and Fokker-Planck Equations

7.1 Distribution Functions and Particle Flux

To follow the details of the cooling process, we may want to know more than just the evaluation of the mean-square beam size and momentum spread—the only quantities used up to now to characterise cooling. In fact, a beam profile monitor records the particle distribution with respect to transverse position (see Fig. 7.1 as an example), and a longitudinal Schottky scan such as Fig. 1.2 gives the (square root of the) momentum distribution. These pictures are rich in fine information on peak densities, densities in the tails, asymmetries, and other practical details which are overlooked if only the r.m.s.-width is regarded.

It is therefore challenging to find an equation which describes the evolution of the particle density distribution. Such an equation does in fact exist!

For stochastic cooling the problem was (to my knowledge) first tackled by Thorndahl [21, 22] who already in 1976 worked with a Fokker-Planck type of equation for the particle density. This line was followed by virtually all subsequent workers [19, 40, 53–55], and computer codes for solving the distribution function equations are extensively used in the design of stochastic cooling and stacking systems. The basic ideas behind this 'distribution function analysis' are simple, so that also the beginner can get some first degree of familiarity with this powerful tool of cooling theory. I will first give the recipe and then try to justify it.

Let $\psi(x)$ (Fig. 7.2) be the particle distribution with respect to the error x (e.g. $x = \Delta p/p$). Define $\psi(x) = dN/dx$ so that it gives the number of particles with an error in the range x to $x + dx$. During cooling we find different distributions. $\psi(x)$, taking snapshots at different times (see Fig. 7.1 as an example). We characterise this by letting $\psi(x, t)$ be a function of time also. The partial differential equation which describes the dynamics of $\psi(x, t)$ can be written in the following form:

$$\frac{\partial \psi}{\partial t} = \frac{\partial}{\partial x}\left(-F\psi + D\frac{\partial \psi}{\partial x}\right) \tag{7.1}$$

The cooling process is completely characterised by the two coefficients F and D (which describe the cooling system) and the initial conditions $\psi(x, 0)$ (which de-

D. Möhl, *Stochastic Cooling of Particle Beams*, Lecture Notes in Physics 866, DOI 10.1007/978-3-642-34979-9_7, © Springer-Verlag Berlin Heidelberg 2013

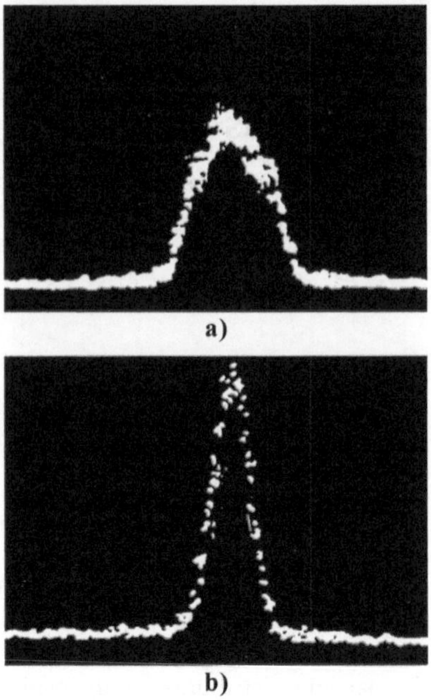

Fig. 7.1 Evolution of beam profile (number of particles vs. vertical position) during stochastic cooling in 'ICE'. The scans were obtained with a profile monitor that records the position of electrons liberated by beam particles through collisions with the residual gas. (**a**) before cooling; (**b**) after 4 min of stochastic cooling (By courtesy of CERN, (©) CERN)

a)

b)

Fig. 7.2 A particle distribution function $\psi(x)$ defining the number of particles $dN = \psi(x)dx$ with an error in the interval from x to $x + dx$ (By courtesy of CERN, (©) CERN)

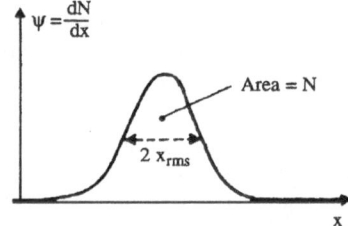

scribe the distribution at the start). Particle loss due to walls or influx during stacking can be included via appropriate boundary conditions $\psi(x_1) = 0$, $(\partial\psi/\partial x)(x_1) = $ const., etc. Two representative examples of results obtained with Eq. (7.1) are given in Fig. 7.3, taken from Refs. [8, 9], and Fig. 7.4 from Ref. [56].

To analyse a given system we have to find its coefficient F and D. These quantities are closely related to the coherent and incoherent effect, respectively, which we have identified before. In fact

$$F/f_{rev} = \langle \Delta x \rangle \tag{7.2}$$

is the expectation value (long-term average) of the coherent change Δx per turn of the error, and

$$2D/f_{rev} = \langle (\Delta x)^2 \rangle \tag{7.3}$$

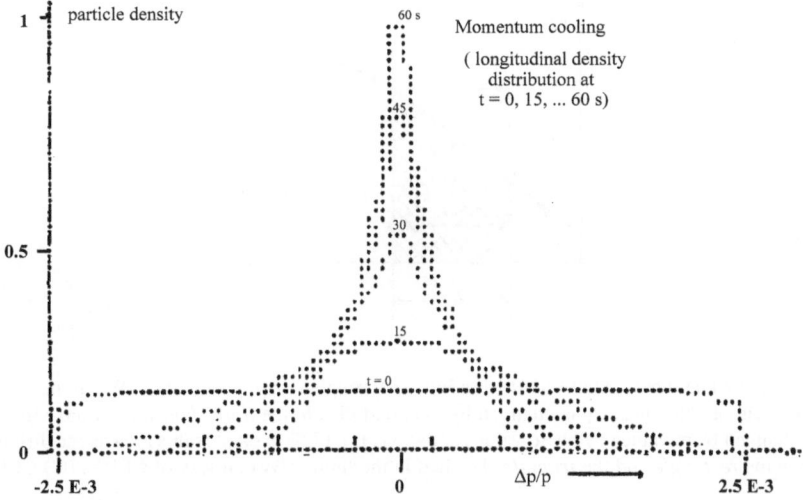

Fig. 7.3 Momentum cooling at 600 MeV/c in LEAR computed using Eq. (7.1) (Curves taken from Refs. [8, 9]) (By courtesy of CERN, (©) CERN)

Fig. 7.4 Evolution of the stack in the AA during stochastic accumulation. Curves computed using the distribution function equation with the boundary condition of constant particle influx simulating the new $\bar{p}$ added every 2.6 s (Curves taken from Ref. [56]) (By courtesy of CERN, (©) CERN)

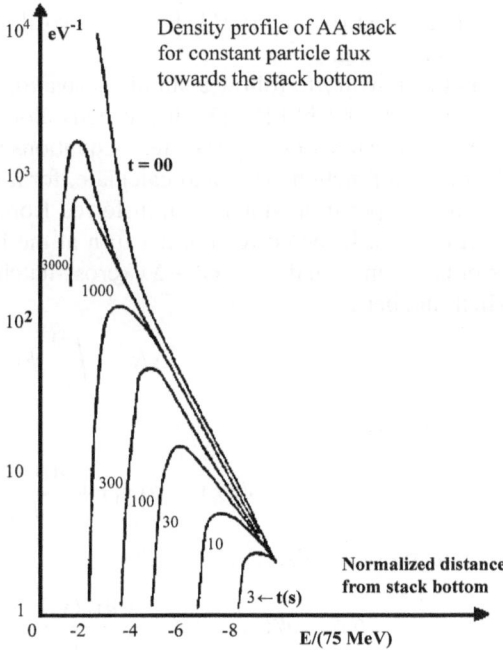

is the expectation of the square of this change. The quantities F and $2D$ alone are corresponding average changes per second.

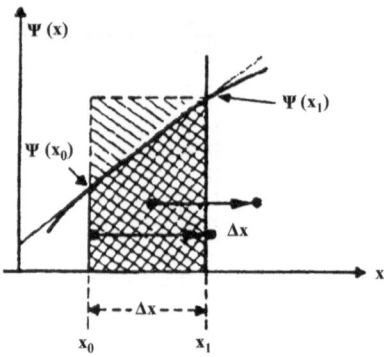

Fig. 7.5 A look at the distribution function Fig. 53 through a magnifying glass. When the error for particles with a value near x_1 is changed by Δx, particles in the *dark shaded area* have the error value changed from values below to values above x_1, Eq. (7.5) expresses this area as the difference between the *rectangle* and the *triangle* sketched in the figure (By courtesy of CERN, (©) CERN)

The important thing is that a distribution function equation—similar to the Fokker-Planck equation used in a variety of fields—exists and that relatively simple prescriptions (7.2) and (7.3) permit us to establish the two coefficients F and D for any given stochastic cooling system. Incidentally, an equation similar to Eq. (7.1) had been used (before 1976!) by the Novosibirsk Group to study the dynamics of electron cooling.

Let us now try to follow a simple derivation of Eqs. (7.1)–(7.3). This derivation is due to Thorndahl [21, 22]. It proceeds along the lines used in textbooks to derive the diffusion—or heat transfer—equations which resemble Eq. (7.1). Imagine a distribution function $\psi(x)$ and calculate, for a particular value x_1 of x the number of particles per turn which are transferred from x-values below x_1 to values above x_1 (Fig. 7.5). If the correction per turn at the kicker is Δx, then particles with an error between x_1 and $x_0 = x_1 - \Delta x$ (cross-hatched area in Fig. 7.3) pass through x_1. Their number is

$$\Delta N = \int_{x_0}^{x_1} \psi(x)dx \tag{7.4}$$

Expanding at x_1

$$\psi(x) \approx \psi(x_1) + \frac{\partial \psi(x_1)}{\partial x_1} \cdot (x - x_1)$$

the integration yields:

$$\Delta N \approx \psi(x_1) \cdot \Delta x - \frac{1}{2}\frac{\partial \psi(x_1)}{\partial x_1} \cdot (\Delta x)^2, \quad \Delta x = x_1 - x_0 \tag{7.5}$$

The first and second term are the area of the rectangle and the triangle, respectively, sketched in Fig. 7.5. We now define the (average) particle flux

$$\phi = f_{rev}\langle \Delta N \rangle$$

as the expected number of particles per second passing a given error value. Clearly, then, from Eq. (7.5), the instantaneous flux is:

$$\phi(x) = \underbrace{f_{rev}\langle\Delta x\rangle}_{F} \cdot \psi(x) - \underbrace{\frac{f_{rev}}{2}\langle(\Delta x)^2\rangle}_{D} \cdot \frac{\partial\psi(x)}{\partial x} \tag{7.6}$$

This gives the flux in terms of F and D as defined by Eqs. (7.2) and (7.3). The assumption has tacitly been made that the change Δx per turn at the kicker is small and $\psi(x)$ smooth, so that higher expansion terms in Eq. (7.5) can be neglected.

Having found the flux we can immediately obtain Eq. (7.1) from the continuity equation:

$$\frac{\partial\phi}{\partial x} + \frac{\partial\psi}{\partial t} = 0 \tag{7.7}$$

It states that the change per second of the density is given by the 'gradient' $\partial\phi/\partial x$ of the flux. This is similar to continuity considerations in other fields like, for instance, the

charge conservation law of electrodynamics:

$$\frac{\partial j}{\partial x} + \frac{\partial\rho}{\partial t} = 0$$

relating current density j and charge density p.

Like other continuity equations, Eq. (7.7) can be obtained by looking at the flux going into and coming out of an element of width dx in e, x-space, as illustrated by Fig. 7.6:

Incoming flux per second: ϕ_1

Outgoing flux per second: $\phi_2 = \phi_1 + \dfrac{\partial\phi}{\partial x}dx$

Surplus per second: $\Delta\phi = \phi_2 - \phi_1 = -\dfrac{\partial\phi}{\partial x}dx$

The resulting density increase (per second) in the element is thus

$$\frac{\Delta\phi}{dx} = -\frac{\partial\phi}{\partial x}$$

and conservation of the particle number requires a $\partial\psi/\partial t$ equal to this.

This completes the derivation of Eq. (7.1). The equation agrees with observations made in the ISR and all subsequent machines using stochastic cooling. The exact form of Eq. (7.1) has been a subject of discussion for some time. Looking at the derivation of the Fokker-Planck equation in textbooks [57, 58], one is tempted to put the coefficient D under the second derivative as is correct for a variety of other stochastic processes. In 1977 a machine experiment [59] was performed at the ISR to clear up this question for cooling and diffusion problems in storage rings. The experiment clearly indicated that in the present case the diffusion term should be $(\partial/\partial x)[D \cdot (\partial\psi/\partial x)]$ in Eq. (7.1) and not $(\partial^2/\partial x^2)[D \cdot \psi]$.

Fig. 7.6 The flux into and out of a narrow element of width dx in ψ, x-space. An excess of incoming over outgoing flux leads to an increase with time of the density $\psi = \Delta N/dx$ of particles in the element (By courtesy of CERN, (©) CERN)

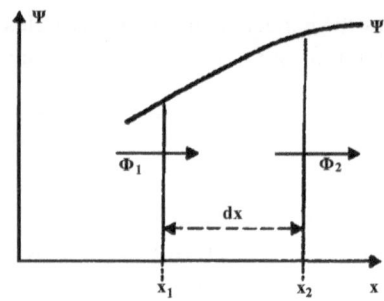

7.2 Simple Example and Asymptotic Distributions

We may conclude from the preceding sections that it is relatively simple to determine the distribution equation pertaining to a given cooling problem. It is usually much more difficult to solve the equation. This is because in general the coefficients F and D are functions of x, t, and ψ itself. Analytical solutions have therefore only been obtained in a few simple cases. As an example, let us briefly look at Palmer cooling with the following simplifying assumption: No unwanted mixing, and Schottky noise negligible compared with amplifier noise.

Denoting $x = \Delta p/p$, the correction per turn is

$$\Delta x = -g\big[\langle x\rangle_s + x_n\big]$$

as given by Eq. (2.24) in Chap. 2. In analogy to Eq. (2.18) in Sect. 2.3, we assume that the long-term sample average; $\langle x\rangle_s = (1/N_s)\sum x_i$ is zero except for the contribution x/N_s of the test particle upon itself. The noise has zero average. Hence

$$\Delta x = -g\frac{1}{N_s}x = -g\frac{2W}{Nf_{rev}}x$$

(where in the last step relation (2.2) between sample and beam population is used).

In a similar way (using the assumption that electronic noise dominates over Schottky noise) we have

$$\big\langle(\Delta x)^2\big\rangle = g^2\langle x_n^2\rangle = g^2 x_{n,rms}^2$$

Hence in this simple case $F = F_0'x$ and $D = D_0$ where $F_0' = g(2W/N)$ and $D_0 = f_{rev}g^2 x_{n,rms}^2$, are constants. In this case, Eq. (4.1) is amenable to an analytic solution. Try

$$\psi = \frac{N}{\sqrt{2\pi}\,\sigma(t)}\,\exp\big[-x^2/\big(2\sigma^2(t)\big)\big]$$

i.e. a Gaussian with a standard deviation σ changing in time. Upon substitution, one obtains an ordinary differential equation for $\sigma(t)$:

$$\dot{\sigma}/\sigma = -F_0' + D_0/\sigma^2$$

The solution is

$$\sigma^2 = \sigma_1^2 \exp(-2F_0't) + D_0/F_0'$$

This describes cooling towards an asymptotic (Gaussian) distribution with $\sigma_\infty = \sqrt{D_0/F_0'}$. In this situation equilibrium between heating and cooling is reached. A similar result is arrived at from the simple cooling equations [e.g. Eq. (2.5), Chap. 2] which suggest $1/\tau \to 0$ when the signal has decreased so much that $gU \to 2$. The new information obtained from Eq. (7.1) is that the asymptotic ψ is Gaussian in the simple case considered.

The existence of asymptotic equilibrium distribution is a common feature also in more complicated cases of Eq. (7.1). The final distribution ψ_∞ can be obtained putting $\partial\psi/\partial t = 0$ which converts Eq. (7.1) into a simpler ordinary differential equation:

$$-F\psi_\infty + D\frac{d\psi_\infty}{dx} = \text{const} \tag{7.8}$$

The constant is frequently zero (e.g. when $F(x) = 0$ and $\partial\psi/\partial x = 0$ for $x = 0$) as can often be inferred from the symmetry of the problem. Equation (7.8) indicates the limiting density which can be reached.

7.3 Calculation of the Coefficients F and D for Filter Cooling

It is customary [19] to use the (total) energy E (in eV) instead of $\Delta p/p$ or ω_{rev} as independent variable. The corresponding distribution functions are related through

$$\psi = \frac{dN}{dE} = \kappa_{\Delta p/p} \cdot \frac{dN}{d(\Delta p/p)} = \kappa \cdot \frac{dN}{d\omega_{rev}}$$

$$\kappa_{\Delta p/p} = \frac{1}{\beta^2 E}, \quad \kappa = \frac{\omega_{rev}\eta}{\beta^2 E} \tag{7.9}$$

In the calculation of F and D we shall assume an effective number n_p of pickups and n_k of kickers, each with a transfer impedance $Z_{P//}$ for the pickups and a voltage transfer function $K_{//}$ for the kickers. The effective number is somewhat smaller than the physical number due to inefficiencies in the combiners. The transfer functions contain both the geometrical sensitivity and the frequency response. For loop couplers their values are tabulated in Table 3.3, Sect. 3.4.

We further assume that the pickups and kickers are matched to the amplifier input and output, and that the complex voltage gain of the cooling loop, including the filters and amplifiers is G'. The total gain from the beam current at the pickup to the kicker voltage seen by the beam is then

$$G'Z_{P//}K_{//}\sqrt{n_p n_k} \tag{7.10}$$

Increasing n_p will improve the signal-to-noise ratio; increasing n_k will reduce the output power needed. If travelling wave pickups or kickers or more complicated devices are used we have to replace $Z_{P//}\sqrt{n_p} \to Z_{PU//}$ or $K_{//}\sqrt{n_k} \to K_{K//}$ in Eq. (7.10) by the total transfer functions.

Each harmonic of the particle's own signal will change its energy by

$$\Delta E_n = 2ef_{rev}\,\mathrm{Re}\big(G'Z_{P//}K_{//}\sqrt{n_p n_k}\big). \tag{7.11}$$

If we include the beam-feedback effect and sum over all harmonics within the system's pass-band, we find

$$F = 2ef_{rev}^2 \sum_n \mathrm{Re}\big(G'Z_{P//}K_{//}\sqrt{n_p n_k}/(1-S_n)\big) \tag{7.12}$$

Here the open loop functions S_n follow from the considerations in Sect. 6.2 and are given by Eq. (5.29) there.

The incoherent effect consists of two components, due to electronic noise and Schottky noise. For the first one, we take again a noise coefficient v as introduced in Eq. (4.32) Sect. 2.3. We referred the noise to the exit of the pickup and assume constant density over the harmonic. To obtain the output power per harmonic, as seen by the particles, we have to multiply the spectral density, Eq. (4.32), by ω_{rev} and obtain

$$\big\langle V_v^2/(R_0)\big\rangle = e^{-v/10}kT_a f_{rev}\big|G'\sqrt{n_k}K_{//}\big|^2 \tag{7.13}$$

In Eq. (7.13) R_0 is the impedance at the exit of the pickup (after recombination i.e. at the input of the pre-amplifier).

The Schottky noise will cause a mean square kicker voltage density per particle and per harmonic of

$$\big(\langle V_{sc}^2\rangle\big) = 2(ef_{rev})^2\big|G'Z_{p//}K_{//}\sqrt{n_p n_k}\big|^2 \frac{dN}{nd\omega_{rev}} \cdot \omega_{rev}$$

$$= 4\pi e^2 f_{rev}^3 \big|G'Z_{p//}K_{//}\sqrt{n_p n_k}\big|^2 \frac{\psi}{(\kappa \cdot n)} \tag{7.14}$$

Here Eq. (4.8) Sect. 4.1 (multiplied with ω_{rev} to obtain the noise per harmonic) and Eq. (7.9) were used. Combining Eqs. (7.13) and (7.14), taking into account the beam-feedback effect, and summing over all harmonics, we find

$$D = \frac{1}{2}kT_a R_0 f_{rev}^2 \sum_n e^{-v_n/10}\left|\frac{(G'\sqrt{n_k}K_{//})}{(1-S_n)}\right|^2$$

$$+ \psi\,\frac{2\pi e^2 f_{rev}^4}{\kappa} \sum_n \frac{1}{n}\left|\frac{G'Z_{p//}K_{//}\sqrt{n_p n_k}}{(1-S_n)}\right|^2 \tag{7.15}$$

Equation (7.15) specifies the diffusion coefficient due to electronic and Schottky noise where we have assumed that the electronic noise can vary with n but is constant over the width of a harmonic.

The expressions (7.12) and (7.15) are used in the (numerical) solution of the Fokker-Planck equation. The open-loop gain, S_n, and the distribution function, ψ appearing in D, change slowly. As an approximation one can therefore keep them constant for a number of turns before updating their values.

7.4 Stochastic Stacking

An important application of stochastic cooling is the accumulation of many successive batches of particles in a storage ring, each filling a substantial fraction of the available phase volume. Following van der Meer [60], this has been extensively analysed by Fokker-Planck techniques. The approach used for the accumulation of antiprotons is stacking in longitudinal phase space where the injection septum as well as the pickups for the momentum cooling are placed in a region of large dispersion (typically $D = 10$ m or higher). To be able to accumulate the required 10^3–10^5 batches a fraction of less than 10^{-3}–10^{-5} is permitted to be lost from the stack at each new injection. This formidable task is (only) attainable by a substantial separation in $D\Delta p/p$ between core and new batch.

The new batch is transported from the injection orbit and placed adjacent to the stack of preceding batches. Cooling then merges the new batch with the stack and reduces the width, so that space is made for the next batch. The density will become much higher for the stack core than for each single batch and the optimum cooling gain will vary accordingly across the distribution, being much higher at the low-density edge. It is necessary to have the highest possible cooling rate at the edge to make space for the next batch as rapidly as possible.

The filter method is used in combination with properly shaped pickups to adjust the gain as a function of momentum. The use of filters requires non-overlapping Schottky bands (bad mixing). The gain profile should allow a constant flux (equal to the number of particles added per unit time) throughout the distribution, from the energy where the new batch is deposited, towards the density peak of the stack.

A simplified description to model the stacking process has been established in the classical paper [60] by van der Meer. It is based on the assumptions that the voltage on the kicker is exactly in phase with the particle that creates it, and that electronic noise and intra beam scattering have a negligible effect compared to the Schottky noise. It is further assumed that the gain in the cooling band is independent of the harmonic number (rectangular characteristic) and the feedback effect is negligible. While none of these assumptions is strictly fulfilled in real systems, the simplified approach leads to results that can form a starting basis. With the above set of assumptions the coefficients F and D, Eqs. (7.12) and (7.15), may be written as:

$$F = 2ef_{rev}^2 n_l G' Z_{P//} K_{//} \sqrt{n_p n_k} = -f_{rev} V$$

$$D = \psi \frac{2\pi e^2 f_{rev}^4}{\kappa} \ln(n_{max}/n_{min}) \cdot \left(G' Z_{P//} K_{//} \sqrt{n_p n_k}\right)^2 = A V^2 \psi \tag{7.16}$$

with

$V = 2ef_{rev}n_l G' Z_{p//}K_{//}\sqrt{n_p n_k}$: the energy gain per turn,

n_{max}, n_{min}, and $n_\ell = n_{max} - n_{min}$: the maximum and minimum harmonic and the number of harmonics in the cooling band,

$A = \frac{f_{rev}^3 \ln(n_{max}/n_{min})\beta^2 \gamma m_0 c^2}{4W^2|\eta|}$: a constant (strength of the Schottky noise).

Following van der Meer the particle flux $\phi = \partial N / \partial t$, Eq. (7.6), is written as

$$\phi = -\frac{V}{T_{rev}}\psi - AV^2\psi\frac{\partial\psi}{\partial E} \tag{7.17}$$

The key observation of Ref. [60] is the following: Optimum conditions are obtained by requiring

(i) constant particle flux, ϕ_0, towards the stack core, equal to the average number of particles injected per second and independent of E;
(ii) a 'gain profile' $V(E)$ such that the density $\psi(E)$ rises as sharply as possible towards the core, in order to minimise the width of the stacking region and hence the aperture needed.

A solution is found in the following way: we regard the situation near the end of stacking when (almost) stationary conditions prevail i.e. when approximately $\psi(E, t) = \psi(E)$ and hence $\partial\psi/\partial E = d\psi/dE$. Using Eq. (7.17) we express the density gradient as

$$\frac{d\psi}{dE} = -\frac{\phi_0}{AV^2\psi} - \frac{1}{AVT_{rev}} \tag{7.18}$$

This expression is maximised if we chose $V(E)$ such that at each value of E

$$V(E) = -\frac{2\phi_0 T_{rev}}{\psi(E)} \tag{7.19}$$

With this condition, Eq. (7.18) becomes

$$\frac{d\psi}{dE} = \frac{\psi}{4A\phi_0 T_{rev}^2} \equiv -\frac{\psi}{E_d} \tag{7.20}$$

Here the 'characteristic energy width', E_d, is defined as

$$E_d = -4A\phi_0 T_{rev}^2 = -\frac{\ln(n_{max}/n_{min}) \cdot \phi_0\beta^2\gamma m_0 c^2}{T_{rev}W^2|\eta|} \tag{7.21}$$

The solution of Eq. (7.20) is

$$\psi = \psi_i \exp\{|E - E_d|/|E_d|\} \tag{7.22}$$

and the corresponding gain profile, Eq. (7.19)

$$V(E) = -\frac{2\phi_0 T_{rev}}{\psi_i}\exp\{-|E - E_d|/|E_d|\} \tag{7.23}$$

Here E_i and ψ_i are the (total) energy and the density in the low intensity (entrance) region of the stack tail; $N_i/\Delta E_i = N_i/\{E_i\beta^2(\Delta p/p)_i\}$ is given by the intensity and the momentum width of the newly added batch. The width of the stack tail, i.e. the distance over which the density (7.22) rises from ψ_i to ψ_{core} is given by:

$$\Delta E_{tail} = E_d \ln(\psi_{core}/\psi_i)$$

$$\frac{\Delta p}{p}/_{tail} = \beta^{-2}\frac{\Delta E_{tail}}{E_{total}} \tag{7.24}$$

Equations (7.22) to (7.24) are the main results of the simplified analysis. They exhibit a gain profile $V(E)$ that *decreases* exponentially and a density profile $\psi(E)$ that *increases* exponentially from the low intensity region to the core, with the same absolute value of the slope (energy width) E_d. This exponential profile determines the minimum width required for the stack. However the effects, neglected in the simplified analysis, will increase this width. The exponential gain characteristic is realised by installing two or more separate cooling systems which are centred on different regions of the horizontal aperture. In addition to these a system based on Palmer cooling (using the left and right 'core electrodes' in Fig. 7.3) is used to keep the core in place.

An example of the evolution of the stack as calculated for the CERN AA was given already in Fig. 7.2. In this case the new batches are deposited at the right (high energy side) and particles are decelerated to the core of the stack near the low energy side. As a second example we discuss in some detail the system designed for the Recycled Experimental Storage Ring (RESR) of the FAIR project at Darmstadt [61]. Here the particles are accelerated to the stack core. The system of pickups is sketched in Fig. 7.7 and parameters for the RESR are compiled in Table 7.1.

The resulting exponential slope, Eq. (7.21), is $E_d \approx 1$ MeV whereas for the real system $E_{d,real} \approx 2$ MeV is chosen. The density distribution after 2, 10, 100 and 1000 injections is shown in Fig. 7.8. The curves are calculated numerically from the full Fokker-Planck equation, with the coefficients given by Eqs. (7.12) and (7.15) Sect. 7.3 including beam feedback and loop coupler electrodes with geometrical efficiency and frequency functions as given in Chap. 3. A large number of trial runs was necessary, to adjust the parameters of the stacking system. A good ratio of the voltage for the auxiliary and the main electrodes (for the geometry sketched in Fig. 1.1) is around 0.3–0.4. The beam is deposited in the centre of main electrodes. The unwanted mixing (delay of the signals compared to beam travelling time pickup to kicker) is minimised for that same point. The "non-mixing point" for the auxiliary system is chosen to be in centre of the auxiliary electrodes and that of the core system in the centre of the gap between right and left core electrodes. Only small additional advantages were found by further fine-adjustment of these deposit and non-mixing positions.

A periodic filter with two notches per harmonic in the main and the auxiliary tail systems (with identical characteristics) reduces excitation of the core by the tail systems. The notches are tuned to frequencies close to the core frequencies (with a small deviation corresponding to $\Delta E = \pm 0.75$ MeV i.e. $\Delta p/p = \pm 2 \times 10^{-4}$, which

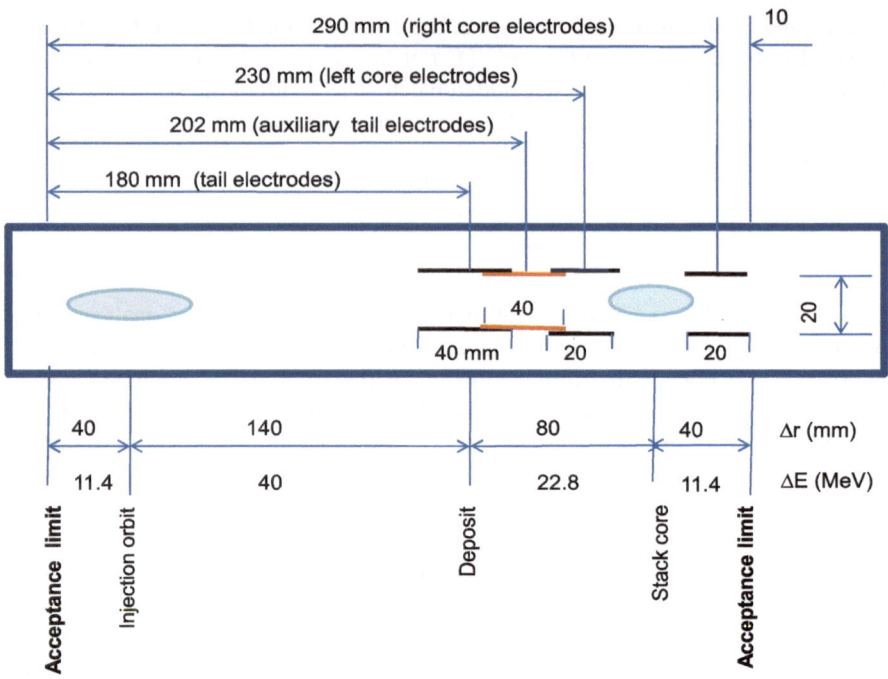

Fig. 7.7 Sketch of a stacking system (radial cross-section at $D = 13$ m) proposed for the RESR. Note that the auxiliary tail electrodes, which partially overlap in horizontal position with the main tail and the left core electrodes, are at different azimuthal positions

amounts roughly to the width of the dense part of the core). The notches have a depth max/min of 1.8/0.2, modelled by a transfer function which is typically

$$F_1 = 0.25 \cdot \left(1 - 0.8 \cdot \exp(i\,\Delta\omega_1/f_1)\right) \cdot \left(1 - 0.8 \cdot \exp(i\,\Delta\omega_2/f_2)\right) \cdot \exp(1.3i)$$

for the main tail system. For the auxiliary system it is:

$$F_2 = 0.25 \cdot \left(1 - 0.8 \cdot \exp(i\,\Delta\omega_1/f_1)\right) \cdot \left(1 - 0.8 \cdot \exp(i\,\Delta\omega_2/f_2)\right) \cdot \exp(2.2i)$$

Here f_1 and f_2 are the frequencies of the minima (the notches) divided by harmonic number and $\Delta\omega_{1,2} = 2\pi(f - hf_{1,2})$ is determined by the frequency deviation from $2\pi h f_{1,2}$. Suitable values for delays (1.3i and 2.2i respectively) and for the depth (0.8) and the positions ($f_{1,2}$) were determined in a series of trial runs.

The stacking systems for the CERN AA and Fermilab antiproton accumulator were designed according to similar considerations, although they use a different number of auxiliary systems and have also somewhat more involved stack protection systems. In fact the original CERN AA and the FERMILAB AA use two auxiliary sets of tail electrodes whereas the upgraded CERN AA, using a 1–2 GHz freq. band instead of the previous 0.15–0.5 GHz band, had only the main set to approximate the ideal gain characteristic.

Table 7.1 RESR beam accumulation parameters

Ring	
Circumference C [m]	249.9
Energy (total) [GeV]	3.94
$\eta = /1/\gamma_{tr}^2 - 1/\gamma^2/$	0.0315
Dispersion at cooling P.U. [m]	13
Incoming beam	
Number of antiprotons/pulse injected $N_{\bar{p}inj}$ [10^8]	1.0
Time between pulses t_{rep} [sec]	10
Momentum spread, at deposit region $\Delta p/p$ (2σ) [10^{-3}]	1.0
Incoming flux $\phi_0(\bar{p}/s)$ [10^7]	1.0
Theoretical (steepest) gain decrement E_d [MeV]	1.0
Adopted gain decrement $E_{d,real}$ [MeV]	2.0
Stack	
Final intensity N [10^{11}]	1.0
Core momentum spread $\Delta p/p$, 2σ [10^{-3}]	0.5
Number of pulses stacked	1 000
Cooling system	
Tail cooling system band [GHz]	1–2
Core cooling system band [GHz]	2–4
Number of 25 Ω tail cooling pickup units, main tail electrodes	64
Number of 25 Ω tail cooling pickup units, auxiliary electrodes	22
Number of 25 Ω core cooling pickup units	64
Number of 100 Ω tail kickers (same kicker at $D = 0$ for main and aux.)	16
Number of 100 Ω core kickers (at $D = 0$)	16
Voltage gain of main tail system	10^7
Voltage gain of auxiliary tail system	3×10^6
Voltage gain of core system	1.5×10^4

7.5   Calculation of the Coefficients F and D for Transverse Cooling

The Fokker-Planck coefficients for transverse cooling can be directly derived from the "coherent" and the "incoherent effect" discussed in Sects. 5.3 and 5.7 respectively. We obtain from Eq. (5.22)

$$F \equiv d\langle \tilde{x} \rangle_c / dt = \frac{x}{2\omega_\beta} \cdot \sum_{m=m_{min}}^{m_{max}} \{ |G(\omega_{m+q})| \sin(\mu - \varphi_{m+q})$$

$$+ |G(\omega_{m-q})| \sin(\mu - \varphi_{m-q}) \} \qquad (7.25)$$

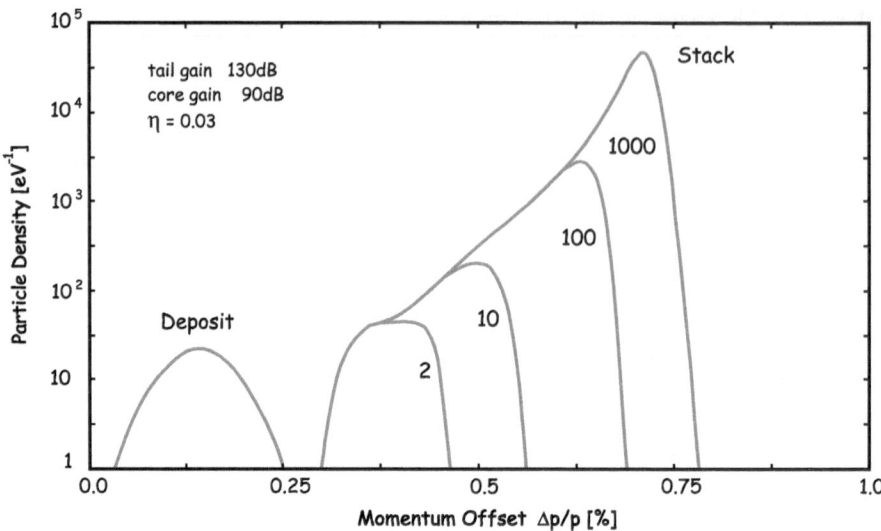

Fig. 7.8 Stacking in the planned RESR of 10^{11} antiprotons from batches of 10^8. Auxiliary electrodes (Fig. 7.7) powered to 30 % of the voltage gain of the main electrodes! Density profiles of new batch, and of stack after 2, 10, 100, 1000 injections, 10 s cycle. Voltage amplification of 10^7 ($\sim$140 db) and 1.5×10^4 ($\sim$84 db) in the tail and core system respectively. The drawing window starts ($\Delta E = 0$) at the left edge of the main tail electrode (Fig. 7.7) and finishes ($\Delta E = 40$ MeV) at the outer edge of the right core electrode

where we have written down the $(m + q)\omega_{rev}$ and $(m - q)\omega_{rev}$ components separately. From Eq. (5.44) we obtain:

$$D \equiv \frac{1}{2} \cdot d\langle \tilde{x}^2 \rangle_{ic} / dt$$

$$= \frac{N x_{rms}^2}{4\omega_\beta^2 f_{rev}} \sum_{m=n_{min}}^{m_{max}} \left\{ \left| G(\omega_{m+q}) \right|^2 \left[M(\omega_{m+q}) + U(\omega_{m+q}) \right] \right.$$

$$\left. + \left| G(\omega_{m-q}) \right|^2 \left[M(\omega_{m-q}) + U(\omega_{m-q}) \right] \right\} \tag{7.26}$$

The above coefficients permit to calculate the evolution of the distribution with respect to the betatron amplitudes.

To include the beam feedback effect one has to calculate (from longitudinal cooling) or assume the longitudinal particle distribution $n(\omega_{l\pm q})$ with respect to the sideband frequencies. The shielding functions $T(\omega_{l\pm q})$ are then obtainable from Eq. (6.13); the harmonics of F have to be multiplied by $T(\omega_{l\pm q})$ and the harmonics of D by $[T(\omega_{l\pm q})]^2$.

Chapter 8
Other Special Applications

8.1 Cooling of Bunched Beams

There is a wide interest in stochastic cooling of bunched beams. The main motivations are: For large hadron colliders:

- to gain in luminosity by reducing the beam size.
- to counteract the luminosity decrease due to emittance degradation.
- to clean the beam halo.

For lower energy machines:

- to facilitate an accumulation method where the stack is concentrated in a (normal or barrier-) bucket over part of the circumference and the new batch is injected onto the free part.

Bunched beam stochastic cooling is more difficult than cooling of the corresponding number of unbunched particles. From the time-domain point of view, it is clear that with decreasing bunch length the sample population increases and the cooling rate is reduced accordingly. In fact we can estimate an upper limit for the cooling rate regarding the bunch as a part of a coasting beam i.e. replacing $N \rightarrow N_b/B_f$. Here N_b is the number of particles per bunch and $B_f =$ [total length of the bunch /circumference] the bunching factor. Then from Eq. (2.31) in Chap. 2:

$$\frac{1}{\tau} \le \frac{W}{N_b/B_f} \left[\frac{(1 - \tilde{M}^{-2})^2}{M + U} \right] < \frac{W}{N_b/B_f} \tag{8.1}$$

Note that the equation can be used for one or several circulating bunches; one just has to take the length and the particle number for each individual bunch.

Equation (8.1) predicts long cooling times for short bunches. In modern colliders the bunching ratio $1/B_f$ is very high, about 10^4 in the former SPS collider [62] and 10^5 in the LHC [63]. In such cases a very large bandwidth is required to obtain cooling times of say 10 hours to counteract the luminosity degradation. Taking the SPS collider as an example, one finds for $N_b = 10^{10}$ particles per bunch and a bandwidth of 5 GHz, a cooling time of 14 hours assuming the mixing and noise term

D. Möhl, *Stochastic Cooling of Particle Beams*, Lecture Notes in Physics 866,
DOI 10.1007/978-3-642-34979-9_8, © Springer-Verlag Berlin Heidelberg 2013

Table 8.1 Bunch parameter assumed for the SPS-collider

Particles/bunch N_b	10^{10}
Bunching ratio $1/B_f$	5×10^3
Cooling band width W [GHz]	5
Mixing and noise $[(1 - \tilde{M}^{-2})^2/(M + U)]$	1/5
Cooling time constant, Eq. (8.1), τ [h]	14

(the rectangular bracket in Eq. (8.1)) to be 1/5 (Table 8.1). This would be just about acceptable to compensate the luminosity decrease due to intra-beam scattering with a decay time of the order of one day.

In addition to the density effect, the mixing will be worse, especially for particles in the centre of the bucket. The relative speed by which these particles overtake each other will vary sinusoidally and for equal synchrotron frequencies the same particles will meet again and again in the same sample. This may also be seen in the frequency-domain. The revolution frequency of each particle is modulated by the much lower synchrotron frequency ω_s. As a result, each Schottky line splits up into a central line (at frequency $n\omega_{rev}$ for longitudinal signals), accompanied by satellite lines spaced by ω_s. Summing over all lines with revolution harmonic n and satellite number k, we may write for the current of a single particle [64]

$$I = ef_{rev} \sum_n \sum_k J_k(n\Delta\varphi) \exp\{i(n\omega_{rev}t + k\omega_s t + k\varphi_s)\} \tag{8.2}$$

where $\Delta\varphi$ is the peak phase excursion due to the synchrotron motion, measured in terms of revolution frequency. We see that contributions from the same k, but neighbouring n are correlated if $n\Delta\varphi$ is small so that the Bessel function J_k changes slowly with n. This is in contrast with the behaviour of unbunched beams, where $k\omega_s t$ is replaced by the initial phase $n\varphi_0$ and the random φ_0 destroys the coherence. As a result, only a fraction of the bunched beam Schottky bands is useful for cooling except when very high frequencies are employed; the distance in frequency of the useful bands is about equal to the inverse of the bunch duration. Using the bands in between will not contribute much.

In the second place, the width of each satellite band is only $k\Delta\omega_s$, where $\Delta\omega_s$ is the spread in synchrotron frequency. As the total area under a Schottky band is constant it is thus split up into peaks and valleys in the bunched beam case with the peaks higher than for the coasting beam. The increased peak density corresponds to bad mixing in the time-domain. Only for high k will this effect be less serious. Since for $k > n\Delta\varphi$ the function J_k becomes small, this means that high harmonic numbers should be used so that large k contribute to the sum in (8.2).

Both effects are less important for particles outside the buckets. It is therefore possible to cool (and trap) these at more or less normal rates, as has been confirmed experimentally [64]. Also for weak beams, where the amplifier noise predominates, bunching does not affect the cooling.

A third problem of a more practical nature exists [65]. The central bands, for $k = 0$, will add up linearly for all particles, as long as $n\Delta\varphi$ is small compared with 1.

Table 8.2 Bunch current roll-off with frequency and 'cross-over frequency' (assuming 10^{10} particles per bunch) for a Gaussian and a rectangular bunch

Bunch shape	Bunch current roll-off	Cross-over frequency for $N_b = 10^{10}$ (frequency above which $I_{schottky}/I_{buch} > 1$)
Gaussian	$I(f)/2N_b f_{rev} = \exp\{-(\pi f T_b)^2/8\}$	$f > 10/T_b$
Rectangular	$I(f)/2N_b f_{rev} = |\sin(\pi f T_b)|/\pi f T_b$	$f > 4500/T_b$

At lower n values this signal will be much stronger than the Schottky signals from the satellite bands, which are proportional to $\sqrt{N}$. The sensitive amplifier needed for the cooling systems may therefore be saturated. To avoid this, filters producing very steep notches at the revolution harmonics are proposed. In addition high harmonic numbers n must (again!) be chosen so that the systematic bunch-signal at $n f_{rev}$ is sufficiently decreased compared to its low frequency value; the required n depends on the bunch shape.

In fact a circulating bunch presents a current spectrum with equidistant lines at the revolution harmonics $n f_{rev}$. The height of the lines can be readily determined by a Fourier development of the circulating bunch. Up to a roll-off, which starts around the 'bunch frequency'

$$f_b = \frac{1}{T_b} = \frac{\beta \cdot c}{l_b} = \frac{f_{rev}}{B_f}, \tag{8.3}$$

the line height is $I_b = 2N_b f_{rev}$. For the two extremes of a Gaussian bunch (with a bunch duration defined as 4 times the r.m.s value: $T_b = 4\sigma_t$) and a rectangular bunch (of total duration T_b), the roll-off function (envelope of I_n) is compiled in Table 8.2. One notes the very much slower roll-off in case of a rectangular bunch which leads to the considerably higher 'cross-over frequency'. Only above this frequency the residual bunch current becomes smaller than the Schottky noise per band [$I_s = \sqrt{2N_b}ef_{rev}$, cf. Eq. (4.9)]. To avoid the strong unwanted bunch signals one prefers to work at frequencies (much) above the cross-over; therefore the cooling band to be chosen depends strongly on the bunch-shape assumed.

For betatron cooling, the situation is similar with the exception of the third effect. The $k = 0$ bands will now be randomised, because of the different initial betatron phase of the particles. However, a strong signal at the lower longitudinal harmonics still exists if the beam is not exactly centred in the pickup.

With the three effects in mind, cooling systems for the SPS collider [62] and the FERMILAB Tevatron [66, 67] were designed. However at that stage pre-experiments observing Schottky noise revealed a strong "RF-activity" at both machines: very intense signals were observed persisting at frequencies considerably higher than $10 f_b$. In fact these signals extended up to the highest frequencies accessible (about 10 GHz). Their nature was not fully [65] understood but it was suspected that coherent instabilities play a role.

Table 8.3 Properties of the RHIC cavity system used for longitudinal cooling

Cooling band [GHz]	5–8
Bunch length, total l_b [m]	1.5
Bunch frequency $f_b = \beta c / l_b$ [MHz]	200
Distance between bunches ΔT [ns]	100
Number of cavities (spaced in resonance by f_b) in the band	16
Average Q-value of cavities $\langle Q \rangle = \langle f_{resonant} \rangle \Delta T$	600
Frequency width of cavities $\Delta f = f_{resonant} / Q$ [MHz]	10
R/Q of cavities [Ω]	120
Impedance at resonance R [K Ω]	72

Tests to explore this problem were performed in LEAR [68] and in the CERN Antiproton Collector ring AC [69]. In LEAR the frequency spectrum of dense bunches obtained after electron cooling showed a cut-off at $5 f_t$, (decrease to 10^{-4} of the low frequency line height) as long as the intensity was low, $N_b < 10^9$ for 150 ns bunch length. However, at $N_b = 5 \times 10^9$, the cut-off was as high as $45 f_b$, thus indicating a strong intensity dependence of the effect. In the AC, a 50 ns long bunch with $N_b \approx 10^7$ particles could be obtained bunching the cooled beam with the $h = 1$ RF system. When switching off the stochastic cooling, strong coherent lines occurred up to the highest observable frequencies (about 3 GHz). They could be reduced to the noise level when the longitudinal cooling system was reactivated. Both longitudinal and transverse cooling then took over, with a (roughly by B_f, as expected from Eq. (8.2)) slower cooling rate, once the longitudinal system, acting as a damper, had ironed out the coherent lines.

The presence of the undesired coherent lines at very high frequencies has been a major obstacle for the successful operation of bunched-beam cooling at the CERN SPS and the FNAL Tevatron.

Since 2002 a serious effort is under way to develop the technique at the Relativistic Heavy Ion Collider (RIC) at Brookhaven [70]. Several innovative steps were explored during this study: The coherent signals were carefully analysed, both by observations at RHIC and theoretically. An explanation was published (in collaboration with workers from the university of Bayreuth, Germany) [71]) linking the high frequency activity to "solitons" i.e. humps or holes in the beam, which result as stationary solutions of a coupled Vlasow-Poisson equation. Both theory and observation suggest, that the effect is weaker for beams of 10^9 Au^{79+} ions in RHIC than for the 10^{11} p and p̄ in the SPS or the TEVATRON.

Longitudinal and vertical cooling systems were therefore installed in the two RHIC rings. Horizontal cooling is achieved by coupling the horizontal to the vertical betatron oscillation. The main RHIC cooling parameters are given in Table 8.3. A factor of 2 in integrated luminosity has been obtained with cooling in only one ring and another factor of 2 is expected when cooling works in both rings. A problem at the high energy is the power of the final amplifier needed. It is solved at RHIC by using a series of high Q cavities for the kicker. The resonant frequen-

Table 8.4 LHC bunch parameters		
Particles per bunch N_b		10^{11}
Bunch length $l_b(4\sigma)$ [m]		0.3
Bunch frequency $\beta c/l_b$ [MHz]		1
Circumference C [km]		27
Bunching ratio $1/B_f$		10^5

Fig. 8.1 Sketch of an electrode structure sensitive to the beam halo

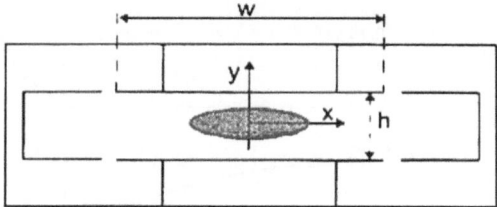

cies are spaced by the bunch frequency $f_b = 1/T_b$ (Eq. (8.3)) and the Q-value is determined by the time ΔT between bunches $Q = f_{resonant} \cdot \Delta T$, so that the filling time of the cavity is ΔT. With the high resonant impedance one arrives at a manageable power requirement (about 20 W/cavity [70]) for the amplifier system.

The cavities have a small beam bore hole matched to the beam size at high energy. During filling and ramping of RHIC they are opened along a vertical mid-plane and then closed for operation during the store. The cavity system for vertical cooling is designed following similar considerations. A transverse deflecting mode is exited rather than the longitudinal field on the cavities for momentum cooling.

8.2 Halo Cleaning

Let us now look at the somewhat different goal of bringing particles from the bunch halo back into the core by some sort of stochastic cooling [65]. In the LHC, this procedure could assist or even partly replace the sophisticated system of collimators to avoid uncontrolled particle loss [63]. A glance at the proton bunch parameters (Table 8.4) convinces us, that stochastic cooling of the entire bunch, as discussed above, at any reasonable timescale is excluded with present-day bandwidth ($W < 10$ GHz).

If however, one could build a pickup which only sees 10^7 particles in the transverse halo instead of the entire beam, then the cooling of these 10^7 particles/bunch might be possible. The problem is to design a pickup with a discrimination core/halo of 10^{-4}. This can perhaps be obtained with a pickup arrangement as sketched in Fig. 8.1.

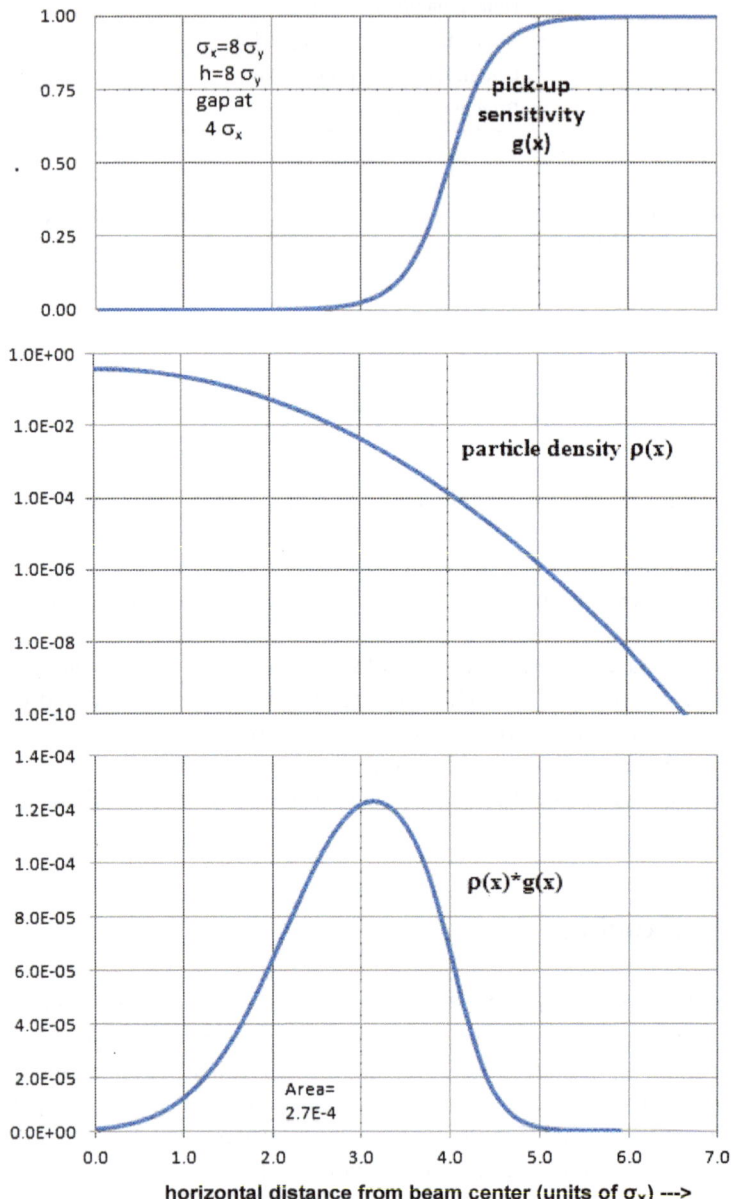

Fig. 8.2 Example of the particle density $\rho(x) \cdot g(x)$ seen near the gap of a tail electrode situated at 4 standard deviations from the centre of a Gaussian beam. The beam is very flat ($\sigma_x = 8\sigma_y$), the vertical aperture is $h/2 = \pm 4\sigma_y$ and $y \approx 0$. The area under the lower curve is 2.7×10^{-4}

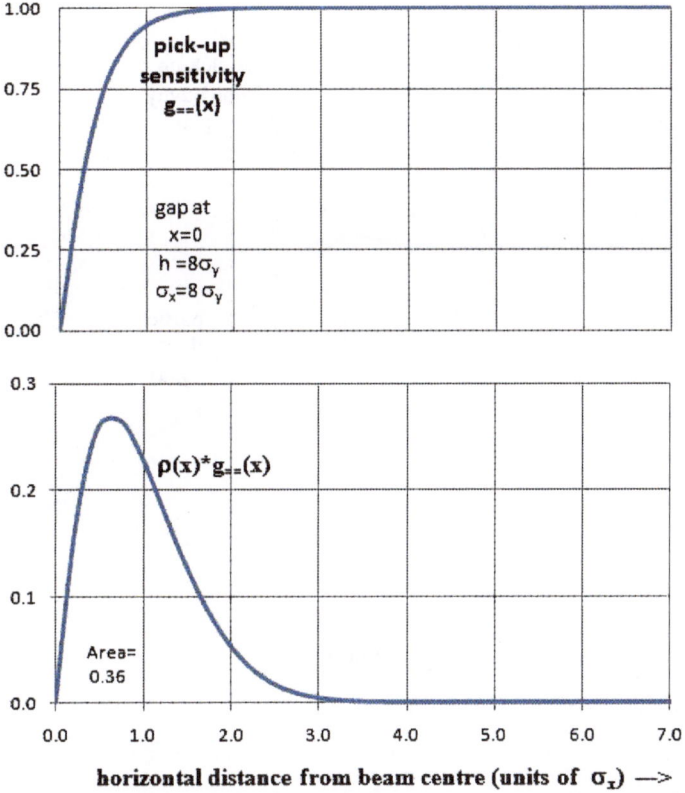

Fig. 8.3 The density detected by a difference electrode (Fig. 3.2) with a small gap centred at $x = 0$ on a Gaussian beam. Parameters ($\sigma_x = 8\sigma_y$, $h = 8\sigma_y$, $y \approx 0$) as in Fig. 8.2. The *curve* continues symmetrically (with negative values) for negative x. The total area under the curve is 0.36 compared to 2.7×10^{-4} for the curve of Fig. 8.2

The sensitivity (for $w \gg h$) at the left and right small gaps can be constructed from the function $g_{\equiv,\equiv}$ given in Table 3.1, Sect. 3.2

$$g = 0.5 + \frac{1}{\pi}\left\{\arctan\left[\frac{\sinh(\frac{\pi}{h}|\Delta x|)}{\cosh(\frac{\pi}{h}y)}\right]\right\} \tag{8.4}$$

Here Δx is the horizontal distance from the gap. Assume a Gaussian beam with the horizontal density function

$$\rho(x) = \frac{1}{\sqrt{2\pi}\sigma}\exp\left(-\frac{x^2}{2\sigma^2}\right) \tag{8.5}$$

Then the right and left gap interact with a fraction $g(\Delta x) \cdot \rho(W/2 + \Delta x)$ of the particles. To obtain a sizable discrimination, one needs a wide electrode ($W \gg h$). The function $g(\Delta x) \cdot \rho(W/2 + \Delta x)$ for an example with the gap at $W/2 = 4\sigma_x$ and

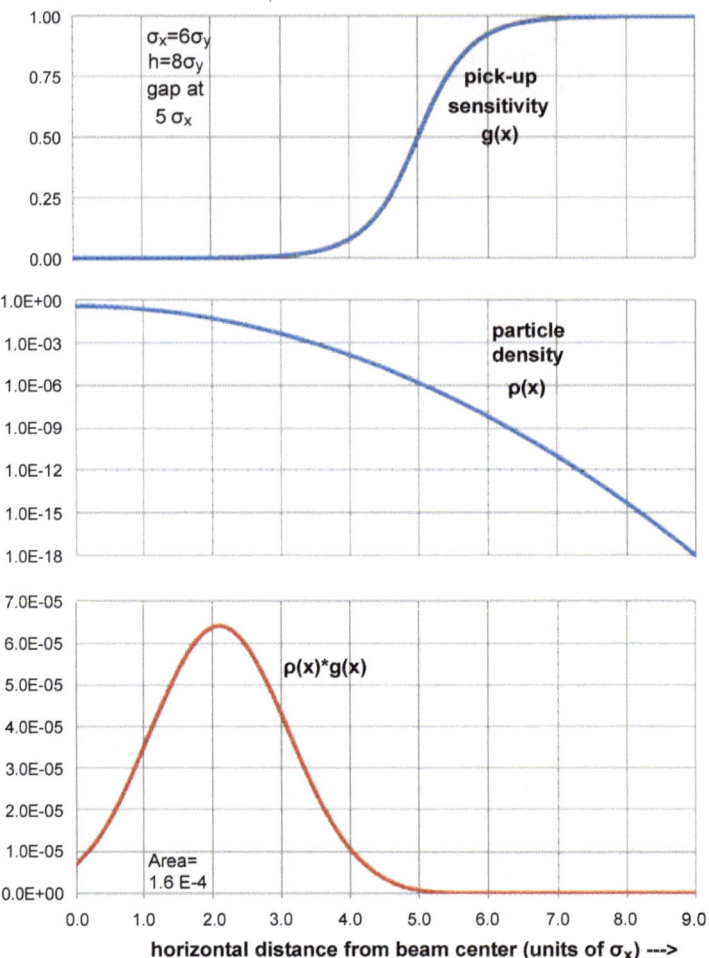

Fig. 8.4 Example of the particle density $\rho(x) \cdot g(x)$ seen near the gap of a tail electrode situated at 5 standard deviations from the centre of a Gaussian beam. The beam is very flat ($\sigma_x = 6\sigma_y$), the vertical aperture is $h/2 = \pm 4\sigma_y$ and $y \approx 0$. The area under the curve is 1.6×10^{-4}

assuming $\sigma_x = 8\sigma_y, h = 8\sigma_y, y = 0$, is shown in Fig. 8.2. This can be compared to the "whole beam situation" observable by a difference pickup (Fig. 3.2, Chap. 3) with a small gap at $x = 0$ (sensitivity $g_{\equiv,\equiv}$ as discussed in Chap. 3, Table 3.1). The relevant function $g_{\equiv,\equiv}(x) \cdot \rho(x)$ is reproduced in Fig. 8.3.

One can conclude, that the ratio (Fig. 8.2/Fig. 8.3) of integrated density is about 3.7×10^{-4} in this example indicating that tail cleaning with an arrangement of Fig. 8.1 may be possible. However a very flat beam is required (with $\sigma_x \approx 8\sigma_y$) which is difficult to realise. Different combinations of parameters (e.g. $W/2 = 5\sigma_x \rightarrow \sigma_x \approx 6\sigma_y, h = 8\sigma_y, y = 0$) lead to a similar discrimination ratio and the resulting beam has the tendency to be flat (see Figs. 8.4 and 8.5).

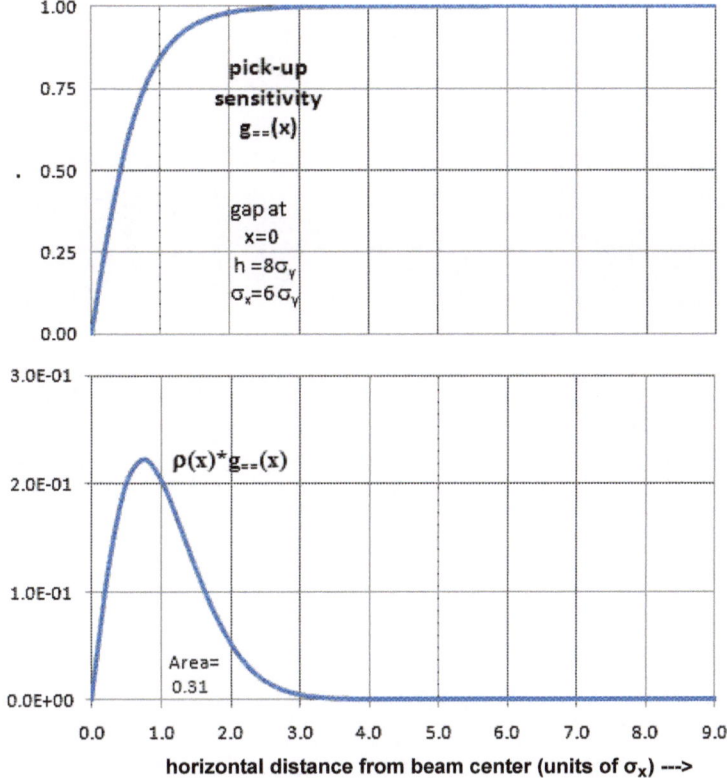

Fig. 8.5 Density detected by a difference electrode (Fig. 3.2) with a small gap centred at $x = 0$ on a Gaussian beam. Parameters ($\sigma_x = 6\sigma_y$, $h = 8\sigma_y$, $y \approx 0$) as in Fig. 8.4. The curve continues symmetrically (with negative values) for negative x. The total area between the *curve* and the x-axis is 0.31 compared to 1.6×10^{-4} for the curve Fig. 8.4

Appendix A
Transfer Function and Power Requirement for Filter Cooling

A.1 Transfer Function

We regard the correlation filter, Fig. 2.15 in Sect. 2.14, and denote by L_1 and $L_2 = L_1 + v_{signal} \cdot T_{rev}$ the length of the signal path from pickup to kicker along the short and the long path respectively. Here T_{rev} is the nominal revolution time and v_{signal} the signal travelling speed. The paths (Figs. 2.14 and 2.15) include the pre-amplifier, the $90°$ phase shifter, the upper and lower part respectively of the correlator, the final amplifiers and all the connecting electronics and cables.

A circulating particle induces a current with harmonics of the form $\cos(n\omega_{rev}t)$ $(= \mathrm{Re}\{e^{in\omega_{rev}t}\})$. We regard the transmission of a single harmonic for the moment. The particle is at the pickup at $t = 0$ and arrives at the kicker at $t = T_{PK}$. The cooling signal takes a time $T_{PK0} = L_1/v_{signal}$ to travel along the short path from pick up to kicker. Along the long path the time is $(T_{PK0} + T_{rev})$. The correction at the kicker at the moment of passage of the particle is the difference between these two contributions

$$S_n = \mathrm{Re}\big[i \cdot \exp\{i\omega(T_{PK} - T_{PKo})\} - i \cdot \exp\{i\omega(T_{PK} - T_{PKo} - T_{rev})\}\big] \quad \text{(A.1)}$$

For a particle of nominal momentum this is zero as $T_{PK} = T_{PK0}$ and $\omega \cdot T_{rev} = 2\pi \cdot n$. For off-momentum particles

$$\begin{aligned} \omega &= n \cdot \omega_{rev0} \cdot (1 - \eta \cdot \delta p/p) \\ T_{PK} &= T_{PK0} \cdot (1 + \eta_{PK} \cdot \delta p/p) \end{aligned} \quad \text{(A.2)}$$

Here $\eta = \gamma_t^{-2} - \gamma^{-2}$ is the usual whole ring slip-factor and η_{PK} is the slip-factor on the way from pickup to kicker. Inserting (A.2) into (A.1) one obtains to first order in $\eta(\delta p/p)$

$$S_n = \sin(n\omega_{rev}T_{PK0}\eta_{PK}\delta p/p) - \sin(n\omega_{rev}T_{PK0}\eta_{PK}\delta p/p + n\omega_{rev}T_{rev}\eta\delta p/p) \quad \text{(A.3)}$$

D. Möhl, *Stochastic Cooling of Particle Beams*, Lecture Notes in Physics 866, DOI 10.1007/978-3-642-34979-9, © Springer-Verlag Berlin Heidelberg 2013

Denoting $\eta_1 = \eta_{PK} \cdot T_{PK0}/T_{rev}$ and $\eta_{eff} = \eta + \eta_1$ and remembering that $\omega_{rev}T_{rev} = 2\pi$, Eq. (A.3) may be written in the form already used in Sect. 2.14

$$S_n = \sin(n \cdot 2\pi \eta_1 \delta p/p) - \sin(n \cdot 2\pi \eta_{eff} \delta p/p) \tag{A.4}$$

This is the response for a single harmonic. For the cooling system we have to add the response of all the harmonics of the pass-band. Figure 2.16 in Sect. 2.14 displays the result for a system with a band-width of one octave and constant (rectangular) gain.

A.2 Power Requirement for Filter Cooling

We regard the set-up sketched in Fig. 2.14 in Sect. 2.14. The beam current power density is the Schottky density, which at the nth harmonic is given by Eq. (4.8) in Sect. 4.1: $d(I)_n^2/d\omega_n = 2(ef_{rev})^2 \cdot dN/d\omega_n$. This current is transmitted to the kicker input impedance through

- the pickup system with transfer impedance $Z_{PU\|}(\omega_n)$
- the amplifiers with a total amplification of $\lambda_a(\omega_n)$
- the filter with a transfer function $S(\omega - \omega_n)$

The filter function depends on the deviation from the centre frequency and thus on the momentum error. The nth harmonic voltage density at the kicker entrance is

$$dV_n^2/d\omega_n = 2(ef_{rev} \cdot Z_{PU\|,n} \cdot \lambda_{a,n})^2 \cdot S_n^2 \cdot dN/d\omega_n \tag{A.5}$$

Taking $Z_{PU\|,n} \cdot \lambda_{a,n}$ constant over the harmonic we find the mean of the squared voltage at harmonic n

$$V_n^2 = 2(ef_{rev}Z_{PU\|,n} \cdot \lambda_{a,n})^2 \int_{(n-1/2)\cdot\omega_{rev}}^{(n+1/2)\cdot\omega_{rev}} N'(\omega_n) \cdot S_n^2(\delta\omega)d\omega_n \tag{A.6}$$

We approximate the function $S_n(\delta\omega) \propto S_n(\delta p/p)$, by a linear function $S_n(\delta\omega) = n \cdot 2\pi \cdot \eta \cdot \delta p/p$, the small-argument expansion of Eq. (A.3). The function $N'(\omega)$, given by the particle distribution, is bell shaped. As an approximation we take a rectangular distribution $N'(\omega) = N/\Delta\omega_n$ of total width $\Delta\omega_n = n \cdot \Delta\omega_{rev} = n \cdot \eta \cdot \Delta p/p$ ["rectangular Schottky bands"]. Then

$$V_n^2 = 2(ef_{rev})^2 N \cdot (Z_{PU\|,n} \cdot \lambda_{a,n})^2 \frac{1}{\Delta\omega_n} \int_{n\cdot\omega_{rev}-\Delta\omega_n/2}^{n\cdot\omega_{rev}+\Delta\omega_n/2} \left[S_n(\delta\omega)\right]^2 d\omega_n$$

$$= \frac{1}{6}(ef_{rev})^2 N \cdot (Z_{PU\|,n} \cdot \lambda_{a,n} \cdot n \cdot 2\pi \cdot \eta \cdot \Delta p/p)^2 \tag{A.7}$$

To calculate the total power we include the electronic noise and sum over the harmonics of the pass-band. The electronic noise of the low level system is filtered by $S(\delta\omega)$ in the same way as the beam signal. Assuming constant noise density given by Eq. (4.32) in Chap. 4, the noise voltage per harmonic at the kicker entrance becomes:

$$V_{v,n}^2 = \frac{1}{2\pi} 10^{v/10} k R T_a \lambda_{a,n}^2 \int_{(n-1/2)\cdot\omega_{rev}}^{(n+1/2)\cdot\omega_{rev}} S_n^2(\delta\omega) d\omega_n \qquad (A.8)$$

With the linear approximation used for $S_n(\delta\omega)$:

$$V_{v,n}^2 = \frac{\pi}{6} 10^{v/10} k R T_a \lambda_{a,n}^2 \omega_r \qquad (A.9)$$

To suppress the noise between (well separated) Schottky bands, one can think of inserting in front of the amplifier a filter with an idealised characteristic

$$|H_n(\omega)| = \begin{cases} 1 & \text{for } |\omega - n\omega_{rev}| < \Delta\omega_n \\ 0 & \text{else} \end{cases} \qquad (A.10)$$

This reduces $V_{v,n}^2$, Eq. (A.9), by a factor $(\Delta\omega_n/\omega_{rev})^3$ but does not influence—in the approximation of rectangular Schottky bands used—the beam signals.

Next we assume that the amplifier has to deliver the power per harmonic

$$P_n = |(V_n^2 + V_{v,n}^2)/R_k| \qquad (A.11)$$

which is independent of the phase shifts involved. The total power is the sum

$$P = \sum_n \frac{V_n^2 + V_{v,n}^2}{R_{k,n}} \qquad (A.12)$$

Here $R_{k,n} = Z_{KK_\|}/K_{KK_\|}$ is the input impedance of the kicker. If n_k kicker units are added via combiners, the total power required reduces by $1/n_k$ compared to the power if only one module is used. Note also that we characterised the filter by the transfer function (A.3). Additional factors due to splitting and recombining are absorbed into the pickup function so that the "single particle cooling rate" is given by

$$\frac{1}{\tau_0} = \frac{1}{\delta p/p} \frac{d(\delta p/p)}{dt}$$

$$= \frac{2e^2 f_{rev}^2}{m_0 c^2 \gamma (\gamma - 1)/(\gamma + 1)} \sum_n Z_{PU\|,n} \cdot K_{KK\|,n} \cdot \lambda_a \cdot \frac{S_n(\delta p/p)}{\delta p/p} \qquad (A.13)$$

In the notation of Chap. 2, the single particle cooling rate is

$$\frac{1}{\tau_0} = \frac{W}{N} 2g$$

Appendix B
Gain and Power Requirement for Palmer System

We regard a Palmer system consisting of

- an array of position sensitive pickups with a global impedance/displacement $Z'_{PU\perp}$,
- an amplifier with the voltage amplification λ_a,
- the connecting cables with probably additional "filters" to shape the response,
- an array of longitudinal kickers with the global voltage transfer function $K_{KK\|}$.

All components are functions of frequency and include phase-shifts, $Z'_{PU\perp}$ and $K_{KK\|}$ also contain geometry factors.

B.1 Gain

For a single harmonic $\omega_n = n\omega_{rev}$ and a single particle circulating with a deviation $\delta p/p$ from nominal momentum, the output at the pickup is $\mathrm{Re}(V_{PU})$ where

$$V_{PU}(\omega_n) = Z'_{PU\perp}(\omega_n) \cdot I_s \cdot D\delta p/p \tag{B.1}$$

Here $I_s = 2ef_{rev}e^{in\omega_{rev}t}$ is the nth harmonic current of the particle [see Eq. (4.1)] and D the value of the dispersion function of the storage ring at the location of the pickup, $x = D\delta p/p$ is the particle displacement due to the momentum error. The voltage experienced by the particle at the kicker is $\mathrm{Re}(V_K)$, where

$$V_K(\omega_n) = \left| Z'_{PU\perp,n} \right| \cdot I_s \cdot D\delta p/p \cdot |\lambda_a| \cdot |K_{KK\|,n}| \cdot e^{-i(\omega_n t + \mu_n)} \tag{B.2}$$

In this equation all quantities are taken at ω_n, μ_n is the sum of all phase-shifts including the time of flight difference ("undesired mixing") pickup to kicker between particle and cooling signal, $i\omega_n(t_{PK} - t_{PK0})$.

D. Möhl, *Stochastic Cooling of Particle Beams*, Lecture Notes in Physics 866, DOI 10.1007/978-3-642-34979-9, © Springer-Verlag Berlin Heidelberg 2013

The voltage (B.2) leads to a change in kinetic energy $\delta E = \mathrm{Re}(e \cdot V_K)$. Thus the change of $\delta p/p$ is

$$
\begin{aligned}
\delta\left(\frac{\delta p}{p}\right) &= \left(\frac{1}{1+\gamma^{-1}}\right)\frac{\delta E}{E} \\
&= \left(\frac{\gamma}{\gamma^2 - 1}\right)\frac{|Z'_{PU\perp,n}| \cdot |I_s| \cdot D \cdot |\lambda_a| \cdot |K_{KK||,n}|}{938MV} \cdot \frac{\delta p}{p} \cdot \mathrm{Re}\left(e^{-i\mu}\right) \quad \text{(B.3)}
\end{aligned}
$$

Equation (4.22) in Sect. 4.2, in the absence of noise may be written as

$$
\delta\left(\frac{\delta p}{p}\right) = \left.\frac{\delta p}{p}\right|_c - \frac{\delta p}{p} = -\lambda_n\frac{\delta p}{p}
$$

Thus $\lambda_n = (2g_n/N) \cdot \mathrm{Re}(e^{-\mu_n})$ [see under Eq. (4.26) in Sect. 4.2] is given by Eq. (B.3) divided by $\delta p/p$:

$$
\lambda_n = \left(\frac{\gamma}{\gamma^2 - 1}\right)\frac{|Z'_{PU\perp,n}| \cdot |I_s| \cdot D \cdot |\lambda_a| \cdot |K_{K||,n}|}{938MV} \cdot \mathrm{Re}\left(e^{-i\mu_n}\right) \quad \text{(B.4)}
$$

Obviously the gain is maximised if the phase μ_n is zero, but this is difficult to obtain and definitely not possible for all harmonics.

For the incoherent effect the phase μ does not matter to the extent that the noise is stationary and does not depend on the arrival time. Then λ_n^2 in the cooling equation (4.26) is determined by the square of Eq. (B.4) without the phase factor. We obtain instead of Eq. (4.28)

$$
\frac{1}{\tau} = \frac{W}{N}\frac{1}{\ell}\sum_{n_1}^{n_1+\ell}\left(2g_n \cdot \cos(\mu_n) - g_n^2 M_n\right) \quad \text{(B.5)}
$$

with

$$
g_n = \frac{N}{2}\left(\frac{\gamma}{\gamma^2 - 1}\right)\frac{|Z'_{PU\perp,n}| \cdot |I_s| \cdot D \cdot |\lambda_a| \cdot |K_{K||,n}|}{938MV} \quad \text{(B.6)}
$$

B.2 Power Requirement

The input current power density is the Schottky noise due to the fluctuation of the particle current, Eq. (4.8) in Chap. 4. Remembering $\Delta\omega/\omega = \eta\Delta p/p$ the voltage power density at the entrance of the kicker is:

$$
dV_n^2/d\omega_n = 2\left[ef_{rev} \cdot |Z'_{PU\perp,n}| \cdot (D/\eta) \cdot (\delta\omega_n/\omega_n) \cdot \lambda_a\right]^2 \cdot dN/d\omega_n \quad \text{(B.7)}
$$

The corresponding noise voltage per harmonic is obtained by integration

$$V_n^2 = 2\left[ef_{rev} \cdot |Z'_{PU\perp,n}| \cdot (D/\eta) \cdot \lambda_{a,n}\right]^2 \int_{(n-1/2)\cdot\omega_{rev}}^{(n+1/2)\cdot\omega_{rev}} \left(\frac{\omega_n - n\omega_{rev}}{n\omega_{rev}}\right)^2 N'(\omega_n) \cdot d\omega_n$$
(B.8)

As an approximation we take a rectangular distribution, $N'(\omega) = dN/d\omega = N/\Delta\omega_n$ of total width $\Delta\omega_n = n \cdot \Delta\omega_{rev} = n \cdot \omega_{rev} \cdot \eta \cdot \Delta p/p$, and a linear function for $(\omega_n - n\omega_{rev})/n\omega_{rev} = \eta\delta p/p$. Then

$$V_n^2 = 2\left[ef_{rev} \cdot |Z'_{PU\perp,n}| \cdot (D/\eta) \cdot \lambda_{a,n}\right]^2 \frac{N}{\Delta\omega_n} \int_{-\Delta\omega_n/2}^{\Delta\omega_n/2} \left(\frac{\omega_n - n\omega_{rev}}{n\omega_{rev}}\right)^2 \cdot d\omega_n$$

$$= \frac{1}{6}\left[ef_{rev} \cdot |Z'_{PU\perp,n}| \cdot D \cdot \lambda_{a,n}\right]^2 \cdot N \cdot \left(\frac{\Delta p}{p}\right)^2$$
(B.9)

To work out the noise to signal ratio we recall that the electronic noise is usually referred to at the entrance of the amplifier. Let us assume, that this noise has the constant power density Eq. (4.32). Then the noise voltage per band referred to at the kicker entrance is

$$V_{\nu,n}^2 = 10^{\nu_n/10} kT_a R_n \lambda_{a,n}^2 f_{rev}.$$
(B.10)

The variables ν, kT_a and R are explained in Sect. 4.3, R_n is the input impedance of the amplifier. To suppress the noise between (well separated) harmonics, one can think of inserting in front of the amplifier a filter with an idealised characteristic

$$|H_n(\omega)| = \begin{cases} 1 & \text{for } |\omega - n\omega_{rev}| < \Delta\omega_n \\ 0 & \text{else} \end{cases}$$
(B.11)

This reduces the electronic noise but does not influence—in the approximation of rectangular Schottky bands used—the beam signals. In fact $V_{\nu,n}^2$ decreases by the factor $\Delta\omega_n/\omega_{rev}$.

We assume that the amplifier has to handle the power per harmonic

$$P_n = \left(|V_n|^2 + |V_{\nu,n}|^2\right)/R_{K,n}.$$
(B.12)

In Eq. (B.12) $R_{K,n} = |\frac{Z_{KK\|,n}}{K_{KK\|,n}}|$ is the total kicker input impedance. When n_k kicker units are combined the power decreases to $1/n_k$ of the value required if only one unit is used.

To obtain the total power, one has to sum (B.12) over the pass-band:

$$P = \sum_n P_n$$
(B.13)

Appendix C
Power Requirement for Transverse Cooling

The total beam Schottky power per harmonic is given by twice the value of Eq. (4.18) as the $n - q$ and the $n + q$ bands contribute:

$$D^2_{schottky} = N e^2 f^2_{rev} \tilde{x}^2_{rms} \tag{C.1}$$

Referred to the kicker entrance the corresponding voltage has to be multiplied by the pickup impedance and the amplification

$$V^2_n = N e^2 f^2_{rev} \tilde{x}^2_{rms} |Z'_{PU}|^2 \lambda^2_{a,n} \tag{C.2}$$

The electronic noise voltage, referred to at the kicker entrance is by virtue of Eq. (4.3)

$$V^2_{v,n} = 10^{v_n/10} k T_a R_n \lambda^2_{a,n} f_{rev} \tag{C.3}$$

The kicker input impedance is $R_{k,n}$ and we take the power as per harmonic $(V^2_n + V^2_{v,n})/R_{k,n}$.

Again we can use relations of Table 3.3 in Sect. 3.4 to express:

$$R_k = \left| \frac{Z_{KK\perp}}{K_{KK\perp}} \right| \text{ or } \left| \frac{Z_{KK==}}{K_{KK==}} \right|$$

depending on the kind of transverse kickers. The total power P is the sum over the $n = W/f_{rev}$ harmonics. If n_k kicker units are combined the power required decreases to $1/n_k$ times the value per unit.

D. Möhl, *Stochastic Cooling of Particle Beams*, Lecture Notes in Physics 866, DOI 10.1007/978-3-642-34979-9, © Springer-Verlag Berlin Heidelberg 2013

Appendix D
Dispersion Integrals

In this appendix we take a brief look at the dispersion integral Eq. (6.8) required for the 'signal shielding' calculations. A more general discussion is given by H.G. Hereward [43] in the context of Landau damping. We need the integral

$$s(\omega) = \int \frac{n(\omega_j)}{\omega_j - \omega} d\omega_j \tag{D.1}$$

contained in Eq. (6.8). It is convenient to work in terms of the deviation from the betatron frequency ω_β denoting $x = \omega_j - \omega_\beta$, $y = \omega - \omega_\beta$. We can write

$$s(y) = \int \frac{n(x)}{(x - y)} dx \tag{D.2}$$

To deal with the singularity of the integrand we assume that the eigenfrequency ω_j of the test particle has a small imaginary part, i.e. we take $\omega_j \to \omega_j + i\alpha$ such that the free oscillation $\tilde{x}^{i\omega_j t - \alpha}$ is damped. Later we go for the limit $\alpha \to 0$.

With the complex eigenfrequency the function $s(x)$ becomes

$$s(x) = \int \frac{n(x) \cdot \{x - y - i\alpha\}}{(x - y)^2 + \alpha^2} dx \tag{D.3}$$

The main contribution to this integral comes from the range $x = y \pm \alpha$ near the pole. For small α we can assume that $n(x)$ is constant in this range and thus take it out of the integral. Integrating from a minimum $\check{x} < y$ to a maximum eigenfrequency $\hat{x} > y$ we obtain for the imaginary part

$$\mathrm{Im}(s) = -\alpha \cdot n(y) \left[\arctan\left(\frac{\hat{x} - y}{\alpha}\right) - \arctan\left(\frac{\check{x} - y}{\alpha}\right) \right]$$

and in the limit $\alpha \to 0$

$$\mathrm{Im}(s) = -n(y) \cdot \pi \tag{D.4}$$

Clearly this is the residuum due to the pole of the integrand. Due to the physics of the problem the value $-i\pi \cdot n(\omega)$ has to be retained.

D. Möhl, *Stochastic Cooling of Particle Beams*, Lecture Notes in Physics 866, DOI 10.1007/978-3-642-34979-9, © Springer-Verlag Berlin Heidelberg 2013

The remaining part of the integral is the principal value. It can be expressed in terms of the Hilpert transform (see Erdelyi et al., Tables of Integral Transforms, vol. 2, McGraw Hill, New York, 1954) defined by

$$H[f(x)] = g(y) = \frac{1}{\pi} \int \frac{f(x)}{x - y} dx \tag{D.5}$$

This transform has been tabulated for a large collection of functions. In terms of the Hilpert transform the principal value of Eq. (D.3) may be written as

$$s_p = \pi \cdot H[n(x)] \tag{D.6}$$

Then

$$s(y) = \pi\{-in(y) + H[n(x)]\} \tag{D.7}$$

Two distributions can serve as models for the construction of approximations like Eq. (6.9) in Sect. 6.1 above:

(1) The 'semi-circular' distribution (which models a distribution with a sharp cut-off)

$$n(x) = \begin{cases} \frac{2}{\pi \Delta^2} \sqrt{\Delta^2 - x^2} & \text{for } |x| \leq \Delta = \Delta \omega_\ell / 2 \\ 0 & \text{elsewhere} \end{cases} \tag{D.8}$$

One obtains

$$s(y) = \frac{2}{\Delta^2}\left\{-i\sqrt{\Delta^2 - y^2} - y\right\} \tag{D.9}$$

(2) The Lorentzian distribution (modeling a distribution with important tails):

$$n(x) = \frac{\Delta/\pi}{\Delta^2 + x^2} \tag{D.10}$$

It yields

$$s(y) = \frac{-i\Delta - y}{\Delta^2 + y^2} \tag{D.11}$$

For the calculation of shielding of the harmonics Eq. (6.13) we need the function

$$S(\omega_\ell) = \frac{N \cdot |G(\omega_\ell)| e^{-i\mu + \varphi_\ell}}{2\omega_\beta} s(\omega - \omega_\ell) \tag{D.12}$$

Appendix E
Longitudinal Beam Response to a Gap Voltage

A voltage $Ve^{i\omega t}$ is applied to a short gap. The voltage experienced by the particles of a coasting beam is $Ve^{i\omega t} \cdot \sum_{-\infty}^{\infty} e^{-in\theta}$ where $\theta = e^{i\omega_{rev}t}$. We retain only the harmonic n with $n\omega_{rev} \approx \omega$, the other harmonics will rapidly get out of synchronism with the particles. The perturbation will modulate the energy, the line density and the revolution frequency of the particles by small contributions, each proportional to $e^{i(\omega t - n\theta)}$.

The energy change per second is

$$\frac{dE}{dt} = eVf_{rev}e^{i(\omega t - n\theta)} \tag{E.1}$$

namely the change $eVe^{i(\omega t - n\theta)}$ per turn times the number f_{rev} of turns per second. Writing

$$E(\theta, t) = E_0 + E_1 e^{i(\omega t - n\theta)}$$

Eq. (E.1) yields Eq. (7.11) of Sect. 7.3 for the energy perturbation E_1/eV

$$\frac{E_1}{eV} = \frac{if_{rev}}{n\omega_{rev} - \omega} \tag{E.2}$$

Next we express the perturbed line density as

$$\lambda(\theta, t) = \lambda_0 + \lambda_1 e^{i(\omega t - n\theta)} \tag{E.3}$$

From the continuity equation

$$\frac{\partial \rho}{\partial t} + \nabla \cdot \vec{j} = 0$$

λ must satisfy

$$\frac{\partial \lambda}{\partial t} + \frac{\partial(\lambda\omega_{rev})}{\partial \theta} = 0 \tag{E.4}$$

D. Möhl, *Stochastic Cooling of Particle Beams*, Lecture Notes in Physics 866, DOI 10.1007/978-3-642-34979-9, © Springer-Verlag Berlin Heidelberg 2013

Suppose the perturbed revolution frequency of the distribution is

$$\omega_{rev}(\theta, t) = \omega_{rev,0} + \omega_1 e^{i(\omega t - n\theta)} \tag{E.5}$$

Then the continuity equation yields:

$$\omega_1 = (\omega - n\omega_{rev,0}) \frac{I_1}{n I_0} \tag{E.6}$$

where $I_0 = R\omega_{rev,0}\lambda_0$, $I_1 \approx R\omega_{rev,0}\lambda_1$ and terms higher than linear in $(\omega - n\omega_{rev})$ have been neglected. We now consider an individual particle in the distribution. The rate of change of its angular frequency is

$$\left(\frac{d\omega_{rev}}{dt}\right) = \frac{\partial\omega_{rev}}{\partial t} + \frac{\partial\omega_{rev}}{\partial\theta}\left(\frac{d\theta}{dt}\right) \approx \frac{\partial\omega_{rev}}{\partial t} + \frac{\partial\omega_{rev}}{\partial\theta}(\omega_{rev,0}) \tag{E.7}$$

where we used the fact that the angular velocity is, to first order, $d\theta/dt = \omega_{rev,0}$. By virtue Eqs. (E.5) and (E.6), Eq. (E.7) is written as

$$\left(\frac{d\omega_{rev}}{dt}\right) \approx i\omega_1 \cdot (\omega - n\omega_{rev,0}) \cdot e^{i(\omega - n\omega_{rev,0})t} = i\frac{I_1}{n I_0}(\omega - n\omega_{rev,0})^2 \cdot e^{i(\omega - n\omega_{rev,0})t} \tag{E.8}$$

Next we take into account the relation between energy and revolution frequency $d\omega/\omega = -(\eta/\beta^2) \cdot dE/E$ from which we have:

$$\left(\frac{d\omega_{rev}}{dt}\right) = \left(\frac{d\omega_{rev}}{dE}\right)\frac{dE}{dt} = \kappa\frac{dE}{dt}, \quad \kappa = -\frac{\eta\omega_{rev}}{\beta^2 E}, \eta = \gamma_{transition}^{-2} - \gamma^{-2} \tag{E.9}$$

Substituting dE/dt from Eq. (E.1) and $d\omega_{rev}/dt$ from Eq. (E.8) and approximating $\omega_{rev} \approx \omega_{rev,0}$ we finally obtain Eq. (7.14) of Sect. 7.3 for the perturbed current

$$\frac{I_1}{eV} = -\frac{if_{rev}n\kappa I_0}{(n\omega_{rev} - \omega)^2} \tag{E.10}$$

Appendix F
Non-linear Pickup and Kicker

The pickup sketched in Fig. 8.1 has a strongly nonlinear characteristic $u(x)$ (upper diagram in Fig. 8.2). We approximate this by a simple function

$$u(x) = S_p x^p \quad \text{with } p \text{ odd} \tag{F.1}$$

Let the corresponding kicker have the characteristic

$$\theta(x) = S_k x^k \quad \text{with } k \text{ even} \tag{F.2}$$

For linear pickup and kicker one has $p = 1, k = 0$.

We estimate the transverse cooling with these nonlinear elements, using time-domain analysis, under a number of simplifying assumptions:

- smooth sinusoidal betatron oscillation
- perfect mixing
- negligible electronic noise
- simplified amplitude distribution

Let the betatron oscillation (in smooth approximation) be given by

$$
\begin{aligned}
x &= \tilde{x} \sin \phi \\
\theta(x) &\equiv dx/d\phi = x' = \tilde{x} \cos \phi
\end{aligned} \tag{F.3}
$$

Here $\phi = Q\omega_{rev}t + \phi_{initial}$ is the betatron phase of the test-particle including a random initial phase, and $\tilde{x}$ is the amplitude. The pickup signal of a test-particle is $u(x)$ (with x the deviation at the pickup). The deflection at the kicker is proportional to $\theta = dx/d\phi = x'$ (with x' taken at the location of the kicker).

A particle passing the cooling system gets its position and angle at the kicker corrected to the new values

$$
\begin{aligned}
x_c &= x \\
x'_c &= x' - \lambda S_k x^k S_p \sum_{sample} (x_i)^p
\end{aligned} \tag{F.4}
$$

D. Möhl, *Stochastic Cooling of Particle Beams*, Lecture Notes in Physics 866, DOI 10.1007/978-3-642-34979-9, © Springer-Verlag Berlin Heidelberg 2013

Here the coordinates (x, x') of the test-particle are to be taken at the kicker; x_i is the deviation of sample member "i" at the pickup, which is situated at a phase advance μ upstream of the kicker; λ, S_k and S_p are constants. We square both equations, and add them to obtain the betatron amplitude $\tilde{x}^2 = x^2 + x'^2$. Using Eqs. (F.3) and (F.4), the change in betatron amplitude becomes:

$$\Delta(\tilde{x}^2) = \tilde{x}_c^2 - \tilde{x}^2$$

$$= -2\lambda S_p S_k \tilde{x}^{k+1} \cos(\phi) \sin^k(\phi) \sum \{\tilde{x}_i \sin(\phi_i - \mu)\}^p$$

$$+ (\lambda S_p S_k)^2 \tilde{x}^{2k} \sin^{2k}(\phi) \left[\sum \{\tilde{x}_i \sin(\phi_i - \mu)\}^p\right]^2 \qquad (F.5)$$

To evaluate the sums, we assume random samples. In the first term of Eq. (F.5), only the contribution of the test-particle to the sum 'coheres' with the part in front, when the average over many revolutions is taken (ϕ uniformly distributed between 0 and 2π). Introducing the notation $g_{pk} \equiv \lambda S_p S_k N_s$, the first term gives

$$\langle \Delta(\tilde{x}_{coh}^2) \rangle = -(2g_{pk}/N_s)\tilde{x}^{k+p+1}\langle \cos(\phi) \sin^k(\phi) \sin^p(\phi - \mu) \rangle \qquad (F.6)$$

The second term averaged over many revolutions gives

$$\langle \Delta(\tilde{x}_{incoh}^2) \rangle = -(g_{kp}/N_s)^2 \tilde{x}^{2k}\left\langle \sin^{2k}(\phi)\left[\sum \{\tilde{x}_i \sin(\phi_i - \mu)\}^p\right]^2\right\rangle \qquad (F.7)$$

This average may be expressed as

$$\left\langle \sin^{2k}(\phi)\left[\sum \{\tilde{x}_i \sin(\phi_i - \mu)\}^p\right]^2\right\rangle = \langle \sin^{2k}(\phi) \rangle\left\langle \left[\sum \{\tilde{x}_i \sin(\phi_i - \mu)\}^p\right]^2\right\rangle,$$

and

$$\left\langle \left[\sum \{\tilde{x}_i \sin(\phi_i - \mu)\}^p\right]^2\right\rangle = \left\langle \sum \tilde{x}_i^{2p} \sin^{2p}(\phi_i - \mu) \right\rangle = N_s \cdot \langle \tilde{x}_i^{2p} \rangle\langle \sin^{2p}(\phi_i - \mu) \rangle$$

Thus collecting terms we finally write the average damping rate per turn:

$$\frac{d\tilde{x}^2}{dn} = -\frac{1}{N_s}\left[2a_{pk}g_{pk}\tilde{x}^{k+p+1} - b_{pk}\langle \tilde{x}_i^{2p} \rangle g_{pk}^2 \tilde{x}^{2k}\right]$$

with $a_{pk} = \langle \cos(\phi) \sin^k(\phi) \sin^p(\phi - \mu) \rangle$ and

$$b_{pk} = \langle \sin^{2k}(\phi) \rangle\langle \sin^{2p}(\phi_i) \rangle \qquad (F.8)$$

Note that the incoherent term depends on the amplitude distribution (via the $\langle \tilde{x}_i^{2p} \rangle$ term).

As in the linear case the sample length is determined by the bandwidth through $T_s = 1/(2W)$ and the sample population, expressed in terms of the bunch population and bunch length, is $N_s \approx N_b T_s / T_b = N_b/(2W T_b)$. In the rest of this appendix, we shall restrict ourselves to a linear kicker ($k = 0$). Then the optimum pickup to kicker spacing is $\mu = \pi/2$ (modulo π) and

$$a_{p0} = \langle \cos^{p+1}(\phi) \rangle = \frac{1}{2^{p+1}} \frac{(p+1)\cdot(p)\cdot(p-1)\cdots(p+3)/2}{1\cdot 2\cdot 3\cdots (p+1)/2}$$

$$b_{p0} = \langle \sin^{2p}(\phi_i) \rangle = \frac{1}{2^{2p}} \frac{(2p)\cdot(2p-1)\cdot(2p-2)\cdots(p+1)}{1\cdot 2\cdot 3\cdots p}$$

(F.9)

$$\frac{d\tilde{x}^2}{dt} = \frac{d\tilde{x}^2}{dn\cdot T_{rev}} = -\frac{2W\cdot T_b}{N_b\cdot T_{rev}}\left[2a_{p0}g_{p0}\tilde{x}^{p+1} - b_{p0}g_{p0}^2\cdot\langle\tilde{x}_i^{2p}\rangle\right]$$

(F.10)

p	1	3	5	7	9	11
a_{p0}	0.5	0.375	0.3125	0.2735	0.2461	0.2256
b_{p0}	0.5	0.3125	0.2461	0.2095	0.1855	0.1682
a_{p0}^2/b_{p0}	0.5	0.45	0.397	0.357	0.326	0.303

We introduce $\tilde{\zeta} = \tilde{x}/\tilde{x}_r$ where $\tilde{x}_r$ is some reference (maximum) amplitude. Then instead of Eq. (F.10):

$$\frac{d\tilde{\zeta}^2}{dt} = -\frac{2W\cdot T_b}{N_b\cdot T_{rev}}\left[2a_{p0}g_{p0}\tilde{x}_r^{p-1}\tilde{\zeta}^{p+1} - b_{p0}g_{p0}^2\tilde{x}_r^{2p-2}\cdot\langle\tilde{\zeta}_i^{2p}\rangle\right]$$

(F.11)

This suggests that the optimum cooling for a particle with amplitude $\tilde{\zeta}$ is characterised by

$$g_{p0} = \frac{a_{p0}}{b_{p0}x_r^{p-1}}\frac{\tilde{\zeta}^{p+1}}{\langle\tilde{\zeta}_i^{2p}\rangle}$$

$$\frac{d\tilde{\zeta}^2}{dt} = -\frac{2W\cdot T_b}{N_b\cdot T_{rev}}\frac{a_{p0}^2}{b}\left[\frac{\tilde{\zeta}^{2p+2}}{\langle\tilde{\zeta}_i^{2p}\rangle}\right]$$

(F.12)

For the reference amplitude the optimum is obtained putting $\tilde{\zeta} = 1$ in Eqs. (F.12)

$$g_{p0} = \frac{a_{p0}}{b_{p0}x_r^{p-1}}\frac{1}{\langle\tilde{\zeta}_i^{2p}\rangle}$$

$$\frac{d\tilde{\zeta}^2}{dt} = -\frac{2W\cdot T_b}{N_b\cdot T_{rev}}\frac{a_{p0}^2}{b_{p0}}\left[\frac{1}{\langle\tilde{\zeta}_i^{2p}\rangle}\right]$$

(F.13)

However at this large gain, small-amplitude particles will be 'anti-cooled' (see Eq. (F.11)) and the centre of the distribution will become depopulated. In fact cooling of the R.M.S. amplitude, obtained by averaging $d\tilde{\zeta}^2/dt$, Eq. (F.11), over the distribution, is:

$$\frac{d\langle\tilde{\zeta}^2\rangle}{dt} = -\frac{2W\cdot T_b}{N_b\cdot T_{rev}}\left[2a_{p0}g_{p0}\tilde{x}_r^{p-1}\langle\tilde{\zeta}^{p+1}\rangle - b_{p0}g_{p0}^2\tilde{x}_r^{2p-2}\cdot\langle\tilde{\zeta}_i^{2p}\rangle\right]$$

(F.14)

The optimum gain and rate for this case are by, respectively, $\langle \tilde{\varsigma}_i^{p+1} \rangle$ and $\langle \tilde{\varsigma}_i^{p+1} \rangle^2$ smaller than the 'optima' for the reference particle.

We shall concentrate on the cooling equation (F.13) for the reference amplitude and evaluate the function $\langle \tilde{\varsigma}_i^{2p} \rangle$ for two different families of distributions. The first:

$$n\big(\tilde{\varsigma}_i^2\big) = \begin{cases} (1+m)(1-\tilde{\varsigma}_i^2)^m & \text{for } 0 < \tilde{\varsigma}_1 < 1 \\ 0 & \text{else} \end{cases} \tag{F.15}$$

We find

$$\langle \tilde{\varsigma}_i^{2p} \rangle = \int_0^1 n\big(\tilde{\varsigma}_i^2\big) \tilde{\varsigma}_i^{2p} d\tilde{\varsigma}_i^2 = \frac{(m+1)!\,p!}{(m+p+1)!}$$

and from Eq. (F.13)

$$\frac{d\tilde{\varsigma}^2}{dt} = -\frac{2W \cdot T_b}{N_b \cdot T_{rev}} \frac{a^2}{b} \left[\frac{(m+p+1)!}{(m+1)!\,p!} \right] \tag{F.16}$$

Especially for $m = 0$ (rectangular distribution) and $m = 1$ (parabolic distribution) we find the optimum for the reference amplitude $\tilde{\varsigma} = 1$:

$$m = 0: \quad \frac{d\tilde{\varsigma}^2}{dt} = -\frac{2W \cdot T_b}{N_b \cdot T_{rev}} \frac{a^2}{b} \left[\frac{p+1}{1} \right] \tag{F.17}$$

$$m = 1: \quad \frac{d\tilde{\varsigma}^2}{dt} = -\frac{2W \cdot T_b}{N_b \cdot T_{rev}} \frac{a^2}{b} \left[\frac{(p+1)(p+2)}{2} \right] \tag{F.18}$$

For a linear pickup, the term in the rectangular bracket in Eq. (F.16) becomes 1 independent of m.

As a second family of distributions we take

$$n\big(\tilde{\varsigma}_i^2\big) = \begin{cases} (1+m)\tilde{\varsigma}_i^{2m} & \text{for } 0 < \tilde{\varsigma}_1 < 1 \\ 0 & \text{else} \end{cases} \tag{F.19}$$

and obtain from Eq. (F.13):

$$\frac{d\tilde{\varsigma}^2}{dt} = -\frac{2W \cdot T_b}{N_b \cdot T_{rev}} \frac{a^2}{b} \left[\frac{p+m+1}{m+1} \right]$$

We conclude that for strong nonlinearity the cooling at large amplitudes can be made stronger than with the linear pickup. This goes however at the expense of an increase of the small amplitudes which will change the distribution. This latter effect can be mitigated by using a nonlinear kicker in addition to the pickup.

Figure Credits

References

1. CERN yellow report 95-06, vol. II, pp. 588–671
2. CERN yellow report 84-15, pp. 97–153
3. CERN yellow report 87-03, vol. II, pp. 453–533
4. Physics Reports **58**, 1980, p. 75

Figure number	From reference
1.1	1
1.2	1
2.1	1
2.2	1
2.3	1
2.4	1
2.5	1
2.6	1
2.7	1
2.8	1
2.9	1
2.10	1
2.11	1
2.12	1
2.13	2
2.14	Original by Author
2.15	Original by Author
2.16	Original by Author
2.17	Original by Author
3.1	Original by Author
3.2	Original by Author
3.3	Original by Author
3.4	Original by Author

D. Möhl, *Stochastic Cooling of Particle Beams*, Lecture Notes in Physics 866, DOI 10.1007/978-3-642-34979-9, © Springer-Verlag Berlin Heidelberg 2013

Figure number	From reference
3.5	Original by Author
3.6	Original by Author
3.7	Original by Author
3.8	Original by Author
4.1	2
4.2	2
4.3	2
4.4	2
4.5	1
4.6	2
4.7	Original by Author
5.1	2
5.2	1
5.3	1
5.4	1
5.5	1
5.6	1
5.7	1
5.8	1
5.9	1
6.1	1
6.2	1
6.3	1
6.4	1
6.5	1
6.6	1
6.7	1
6.8	4
6.9	4
7.1	1
7.2	1
7.3	1
7.4	1
7.5	1
7.6	1
7.7	Original by Author
7.8	Original by Author
8.1	Original by Author
8.2	Original by Author
8.3	Original by Author
8.4	Original by Author
8.5	Original by Author

Glossary

E_h [π rad m]	horizontal emittance
E_v [π rad m]	vertical emittance
$E(x)$	expectation value of x [Eqs. (2.14a)–(2.14c), p. 13]
$E = m_0 c^2 \gamma$ [eV]	energy
p [eV/c]	momentum
N	number of particles in a beam
$N_s = N/(2WT_{rev})$	number of particles per sample (in a coasting beam)
$\Delta p/p$	momentum spread of beam (usually full momentum width, $\approx 4\sigma$)
λ_0 [m]	betatron oscillation wave length
λ, λ_i [m]	fractional change of error of test particle due to its own action and due to the action of particle "i" in the same sample [Eq. (2.5)]
x [m]	error variable of a particle, horizontal or vertical deviation from nominal orbit or from momentum deviation $\delta p/p$
x_c [m]	corrected value of x (after passage through cooling system)
$\tilde{x}$ [m]	amplitude of horizontal betatron oscillation $x(s) \approx \tilde{x} \cdot \sin\{Qs/R + \mu_0\}$ p. 7, Fig. 2.3
x'	angular deviation of orbit $x'(s) \approx \tilde{x} \cdot \cos\{Qs/R + \mu_0\}x$ [p. 7, Fig. 2.3]
$T_{rev} = 1/f_{rev}$ [s]	revolution time of a particle
ψ and $\Delta\psi$ [radian]	betatron phase and betatron phase advance
$\psi = dN/dE$ [(eV)$^{-1}$]	particle distribution in energy (Chap. 7)

D. Möhl, *Stochastic Cooling of Particle Beams*, Lecture Notes in Physics 866,
DOI 10.1007/978-3-642-34979-9, © Springer-Verlag Berlin Heidelberg 2013

Bibliography

1. A.M. Sessler, Methods of beam cooling, in *Proc. Workshop on Crystalline Beams and Related Topics*, ed. by D.M. Maletic, A.G. Ruggiero (World Scientific, Singapore, 1996), p. 93
2. D. Möhl, A.M. Sessler, Beam cooling: principles and achievements. Nucl. Instrum. Methods A **532**, 1 (2004)
3. S.Y. Lee, *Accelerator Physics* (World Scientific, Singapore, 1996)
4. H. Bruck, *Accelerateur Circulaires des Particules* (Bibliotheque des Sciences Nucleaires, Paris, 1966)
5. C.S. Taylor, Stochastic cooling hardware, in *Proc. CERN Acc. School (Antiprotons for Colliding-Beam Facilities)*, Geneva (1983), CERN/84-15, p. 136
6. W. Meyer Eppler, *Grundlagen und Anwendungen der Informationstheorie* (Springer, Berlin, 1966)
7. A. Betts, *Signal Processing and Noise* (English Universities Press, London, 1970)
8. Lear Design Team, Design study of a facility for experiments with low energy antiprotons (LEAR). Int. Report CERN-PS-DL-80-7, 1980
9. P. Lefèvre et al., The low energy antiproton ring (LEAR) project, in *Proc. 11th High Energy Acc. Conf.* (1980), p. 819
10. M.R. Spiegel, *Statistics*, Schaum Outline Series (McGraw-Hill, New York, 1972)
11. L. Maisel, *Probability, Statistics and Random Processes* (Simon and Schuster, New York, 1971)
12. H.G. Hereward, Statistical phenomena—theory, in *Proc. 1st Course of International School of Particle Accelerators*, Geneva (1976), CERN/77-13, p. 281
13. M. Bregman et al., Measurement of the antiproton lifetime using the ICE storage ring. Phys. Lett. B **76**, 1 (1980)
14. E.J.N. Wilson (ed.), Design study of an antiproton collector for the antiproton accumulator (ACOL). CERN/83-10
15. E. Jones, Progress on ACOL, in *Proc. of 3rd LEAR Workshop, Physics with Cooled Low Energy Antiproton Beams*, Tignes (Editions Frotiers, Gif-sur-Yvette, 1985), p. 9
16. D. Möhl, Status of stochastic cooling. Nucl. Instrum. Methods A **391**, 164 (1997)
17. A.A. Mikhailichenko, M.S. Zolotorev, Optical stochastic cooling. Phys. Rev. Lett. **71**, 4146 (1993)
18. M.S. Zolotorev, A.A. Zholents, Transit time method of optical stochastic cooling. Phys. Rev. E **50**, 3084 (1994)
19. D. Möhl, G. Petrucci, L. Thorndahl, S. van der Meer, Physics and technique of stochastic cooling. Phys. Rep. **58**, 75 (1980)
20. S. van der Meer, Stochastic damping of betatron oscillations in the ISR. Int. Report CERN/ISR-PO/72-31

D. Möhl, *Stochastic Cooling of Particle Beams*, Lecture Notes in Physics 866,
DOI 10.1007/978-3-642-34979-9, © Springer-Verlag Berlin Heidelberg 2013

21. l. Thorndahl, A differential equation for stochastic cooling of momentum spread with the filter method. Int. Note CERN/ISR-RF Technical Note LT/ps, 1977
22. G. Carron, L. Thorndahl, Stochastic cooling of momentum spread with filter techniques. Int. Report CERN/ISR-RF78-12
23. W. Kells, Filterless fast momentum cooling, in *Proc. 11th Int. Conf. on High Energy Acc.*, Geneva (Birkhauser, Basel, 1980), p. 777
24. G. Lambertson, Dynamic devices—pick-ups and kickers, in *Proc. US Particle Acc. School 1984/85*. AIP Conf. Proc., vol. 153 (1987), p. 1415
25. R. Shafer, Notation for loop pick-up geometrical sensitivity. FERMILAB Internal Note, $\bar{p}$ Note 232, Sept. 1982
26. F. Caspers, Planar slotline pick-ups and kickers for stochastic cooling. Int. Report CERN/PS/AA 85-48
27. F. Caspers, in *Proc. PAC87*, Washington (1987), p. 1866
28. J. Petter, J. Marriner, D. McGinnis, Novel stochastic cooling pick-ups/kickers, in *Proc. PAC89*, Chicago (1989), p. 636
29. L. Thorndahl, Fixed 90 mm full-aperture structures with TM10-mode propagation for stochastic cooling in HESR. GSI Internal Document, 2006
30. L. Thorndahl, Beam impedances of Faltin structures for stochastic stacking in the RESR. GSI Internal Document, 2008
31. C. Peschke, F. Nolden, Pick-up electrode system for the CR-stochastic cooling system, in *Proc. COOL 2007*, Bad Kreuznach (2007), p. 194
32. C. Peschke, F. Nolden, M. Balk, Planar pick-up electrodes for stochastic cooling. Nucl. Instrum. Methods A **532**, 459 (2004)
33. R. Stassen, Pick-up structures for HESR stochastic cooling system, in *Proc. EPAC2006*, Edinburgh (2006), p. 228
34. R. Stassen et al., Recent developments for the HESR stochastic cooling system, in *Proc. COOL 2007*, Bad Kreuznach (2007), p. 191
35. L. Faltin, Slot-type pick-up and kicker for stochastic beam cooling. Nucl. Instrum. Methods **148**, 449 (1978)
36. J. Borer et al., Non-destructive diagnostic of coasting beams with Schottky noise, in *Proc. IXth Int. Conf. on High Energy Acc.*, Stanford (1974), p. 53
37. S. van der Meer, Diagnostics with Schottky noise, in *Proc. Joint US-CERN School on Particle Acc., Frontiers of Particle Beams*. Lecture Notes on Phys., vol. 343, Capri, 1988 (Springer, Berlin, 1989), p. 423
38. M. Steck et al., in *Proc. Workshop on Electron Cooling and New Cooling Techniques*, ed. by R. Calabrese, L. Tecchio, Legnaro, Italy, 1990 (World Scientific, Singapore, 1991), p. 64
39. S. Chattopadhyay, Some fundamental aspects of fluctuations and coherence in charged-particle beams in storage rings. CERN/84-11
40. F. Sacherer, Stochastic cooling theory. Int. Report CERN-ISR-TH 78-11
41. S. van der Meer, Influence of bad mixing on stochastic acceleration. Int. Report CERN-SPS-DI/pp 77-8
42. D.W. Jordan, P. Smith, *Nonlinear Ordinary Differential Equations* (Oxford University Press, Oxford, 1977)
43. R.M. Howard, *Principles of Random Signal Analysis and Low Noise Design: The Power Spectral Density and Its Applications* (Wiley, New York, 2002)
44. G. Dôme, Theory of rf acceleration, in *Proc. CERN Acc. School*, CERN/87-03, p. 110
45. H.G. Hereward, K. Johnsen, The effect of radio frequency noise. CERN/60-38
46. S. van der Meer, Gain adjustment criterion for betatron cooling in the presence of amplifier noise. Int. Report CERN/PS/AA/ Note 82-2
47. S. van der Meer, Optimum gain and phase for stochastic cooling, in: *Proc. CERN Acc. School, Antiprotons for Colliding Beam Facilities*, CERN/84–15, p. 183
48. H.G. Hereward, The elementary theory of Landau damping. CERN/65-20
49. A. Hofmann, in *Coherent Instabilities*, ed. by M. Dienes, M. Month. Lecture Notes on Phys., vol. 400 (Springer, Berlin, 1990)

50. A. Hofmann, Landau damping, in *Proc. CERN Acc. School*, CERN/2006-02, p. 271
51. A. Hofmann, Single beam collective phenomena-longitudinal, in *Proc. 1st Course of Int. School of Particle Accelerators*, CERN/77-13, p. 139
52. D.A. Edwards, M.J. Syphers, *An Introduction to the Physics of High Energy Accelerators* (Wiley, New York, 1993)
53. L.J. Laslett, Evolution of the amplitude distribution function. LBL-Berkeley Report LBL-6459, 1977
54. J. Bisognano, Kinetic equations for longitudinal stochastic cooling. LBL-Berkeley Report BECON-10, 1979
55. J. Bisognano, in *Proc. 11th Int. Conf. High Energy Acc.*, Geneva (1980), p. 772
56. AA Design Team, Design study of a proton–antiproton colliding facility. Int. Report CERN/PS/AA 78-3, 1978
57. G. Ischimaru, *Basic Principles of Plasma Physics* (Benjamin, Reading, 1973)
58. H. Risken, *The Fokker-Planck Equation* (Springer, Berlin, 1996)
59. L. Thorndahl et al., Diffusion in momentum space caused by filtered noise. Internal Report CERN/ISR-RF-TH, Machine Performance Report 19, August 1977
60. S. van der Meer, Stochastic stacking in the antiproton accumulator. Int. Report CERN/PS/AA 78-22
61. FAIR Baseline Technical Report, March 2006. http://www.nipne.ro/international/cooperations/fair/Executive_Summary.pdf
62. D. Boussard et al., Feasibility study of stochastic cooling of bunches in the SPS, in *Proc. CERN Acc. School (Antiprotons for Colliding-Beam Facilities)*, Geneva (1983), CERN/84-15, p. 197
63. O. Brüning et al., LHC design report (volume 1: the LHC main ring). CERN/2004-003-v-1
64. H. Herr, D. Möhl, Bunched beam stochastic cooling, in *Proc: Workshop on the cooling of high-energy beams*, Madison, USA (1978), p. 41. Preprint CERN-EP-Note-79-34; CERN-PS-DL-Note-79-3
65. F. Caspers, D. Möhl, Stochastic cooling in hadron colliders, in *Proc XVII Int. Conf. on High Energy Acc. (HEACC'98)*, ed. by I. Meshkov, Dubna (1998), p. 398
66. R.J. Pasquinelli, Bunched beam cooling for the Fermilab Tevatron, in *Proc. PAC 95*, Dallas (1995), p. 2379
67. G. Jackson et al., Bunched beam stochastic cooling in the Fermilab Tevatron collider, in *Proc. PAC 93*, Washington (1993), p. 3533
68. J. Bosser et al., LEAR MD report: bunched beam Schottky spectrum. Internal Note, CERN/PS/AR Note 94-15(MD), 1994
69. F. Caspers et al., Bunched beam stochastic cooling, in *Cooling Club Newsletter*, ed. by G. Tranquille. CERN, PS (1994)
70. M. Blaskiewicz, J.M. Brennan, K. Mernick, Three-dimensional cooling in the Relativistic Heavy Ion Collider. Phys. Rev. Lett. **105**, 094801 (2010)
71. M. Blaskiewicz, J. Wei, A. Luque, H. Schamel, Longitudinal solitons in bunched beams. Phys. Rev. Spec. Top., Accel. Beams **7**, 044402 (2004)